INCREDIBLE OTTOMAN PROJECTS

INCREDIBLE OTTOMAN PROJECTS

Turan Şahin

İstanbul/2013

BLUE DOME

INCREDIBLE OTTOMAN PROJECTS

16 15 14 13 1 2 3 4

Editor in Chief
Reşit Haylamaz

Content Editor
Seyit Erkal

Author
Turan Şahin

Editor
Salih Gülen

Art Director
Engin Çiftçi

Graphic Design
Nihat İnce

Page Layout
Mehmet Buma

Translation and Editing
AIM Consulting/ AIM Danışmanlık

ISBN
978-1-935295-27-3

Publication Number
56

Print and Binding
Numune Matbaacılık ve Cilt San.Tic.Ltd.Şti.
Ser. No: 12151
Yüzyıl Mah. Matbaacılar Sitesi
2.Cad No:192 Bağcılar / İST.
Tel: (0212) 629 02 02
www.numune.com.tr
February 2013

CONTENTS

INTRODUCTION

In addition to written and visual documents, every formerly great state leaves artifacts behind. Therefore, any stone, seemingly ordinary brush stroke or a torn piece of paper possibly having witnessed the historical adventure of humanity is ascribed the attribute of "historical artifact." Any ruin or structure, regardless of purpose, which has remained standing, is actually a great historical corpus containing data and secrets. Sometimes a statue, sometimes a grand palace or house of worship, or even a seemingly unimportant painting, may reflect the grandeur of that civilization.

Every civilization, empire and emperor wants to have its power reflected in a lasting work. This historical expression takes the form of cities, places of worship, palaces, monumental public works, statues, monuments and triumphal arches.

At the time when the Ottoman State appeared on the historical stages, it was the practice of great sovereigns to have magnificent works of art done by artists among their subjects. Primarily designed to reflect the grandeur of the rulers, they also embodied elements of their religious faith and geographical location. In the Islamic world, rulers preferred to construct huge houses of worship and monumental mausoleums inscribed with panegyrics praising their achievements, and other enormous public works ensuring that their names and the values for which they stood would live on.

Different idioms were used in the West for the same purpose. While public works and projects were important, statues and paintings were also extremely common. Given Islam's view of painting and statues, Muslim rulers demanding works symbolizing the achievement of the civilization steered the Muslim artist toward architectural projects.

One of those artists serving the Ottoman State was Sinan-i Atik, who justifiably ignored the Sultan's request for an epitomization of the power of the Sultanate. Consequently, when the Fatih Mosque is complete, (Fatih) Mehmet II would make the architect, whom he calls before him, account for his actions: "Why didn't you build my mosque as lofty as Hagia Sophia, but much lower?"[1] In fact, during the opening of the Hagia Sophia, he was so impressed as to say to his builder Justinianus, "O Süleyman! I have surpassed you,"[2], referring to the Prophet Sulayman. Sinan-i Atik's intention was to shame the Great Emperor Justinian, who had demanded a show of power from his architects, who constructed a dome on Hagia Sophia, which collapsed in an earthquake twenty years later and had to be repaired after every subsequent earthquake, and to avoid bringing this shame upon Mehmet II. Nevertheless, unfortunately, the dome that had been kept low to avoid damage in the event of an earthquake would collapse in the 1776 earthquake.

The new rulers of Constantinople did not abandon this desire for grandeur until the state had become outdated. Indeed, the letter that came to Topkapı Palace on 3 July 1503 from "a non-Muslim by the name of Lecardo" was evidence of recourse to this power.[3] Two years after Leonardo da Vinci, Michelangelo also had plans to seek sanctuary in the power of the great empire.[4] Michelangelo, who had had a rift with the Pope, was conducting negotiations through a Franciscan monk by the name of Friar Timoteo, who lived in İstanbul, and receiving information about the reigning sultan and İstanbul. When Michelangelo asked, "Will I be free? Will the sultan allow me to do whatever I want artistically?"

The elderly monk responded: "Listen, my son; during the time of the Greeks, a sculptor wanted to make a sculpture for the city of Alexandria so great that the entire city—its houses, squares and streets—would fit into its palm. If you have the courage to do something on a grander scale than this, only in İstanbul will you be able to do so. The sultan—a thousand times more powerful and richer than the Pope—desires the construction of something the magnificence of which no mortal has yet to imagine."[5]

1 Evliya Çelebi, *Evliya Çelebi Seyahatnamesi;* İstanbul: YKY, 2003, vol.1, p. 101

2 Koçin, Abdülhakim, "Çağını Aşanlar," *Bilim ve Teknik*, Sept. 1991, p. 51

3 A copy of the translation of the letter Leonardo da Vinci sent to Bayezid II is located in Topkapı Palace Archive E 6184

4 Eyice, Semavi, *Belgelerle Türk Tarihi Dergisi*, Issue 19. p. 24

5 İlter, İsmet, *Boğaz Geçişlerinin Tarihçesi*, Karayolları Genel Müdürlüğü Matbaası, 1943, p. 53

6 Necipoğlu, Gülru, *The Age of Sinan*, Hong Kong: Reaktion Books, 2005, p. 209

This is why those who understand Süleymaniye can see it as a uniquely 16th-century Ottoman work of art. However, Süleymaniye cannot be attributed to only Sultan Süleyman or Sinan the Architect. It is a work symbolizing the artistic and financial power of a civilization at its pinnacle.

Süleymaniye, with its magnitude, expresses the strength of the values personally championed by Sultan Süleyman. Visiting İstanbul during the years when it was built, Dersschwan stresses that the Süleymaniye Külliyesi (social complex) is larger than the cities of Bratislava and Belgrade.[6] It appears that aim of the sultan was for the *külliye*, with not only the mosque but also the group of buildings surrounding it, to be larger than many cities in Europe.

There were more than just a few Süleymaniyes in Ottoman history. What they said about Süleymaniye would be said in subsequent years about dozens of works and projects, such as Selimiye, the initiative to merge the Don and Volga, the Sultan Ahmet Mosque (the Blue Mosque), the Tünel (Tunnel), the Hejaz Railway and the Danube-Black Sea Canal. This was the case not only for works that got carried out but also about projects that got no farther than the drawing board and other similar initiatives sharing the same fate.

Let's return to the words of Friar Timoteo: "...only in İstanbul can you do it." He was undoubtedly right; the rulers of this city had the educational and cultural background that enabled them to seek the distinct, the most innovative and the unthought-of. In his work *The Muqaddimah*, Ibn Khaldun says, "The taste of the state and its rulers shapes the artistic activities of that country." Süleyman the Magnificent was after quality, and so he sought Sinan. No work of the master architect, who designed and implemented of 477 works in his long life, resembled another. The capabilities of the client forces the artist to go that much further when working on each new order most of the Ottoman sultans received intensive education during their childhood and were practitioners of a single branch of art. There were more than a few sultans who were poets with collections of poems, jewelers with a high level of design acumen, carpenters, calligraphers and even composers. Their personal capabilities were not limited to a branch of art. These individuals were familiar with different cultural forms and were good actors who could also speak in different languages. They were well abreast of the political, cultural, economic and technological developments of the period in which they lived. Therefore, they made a point of ordering projects that had the symbolic value of representing their positions. It was difficult to convince clients requesting a project who have the background and financial wherewithal to do so. Not every architect, engineer or artist was capable of responding to these requests.

Without a doubt, public demands also played a part in the realization of these projects. Ottoman geography was such that many vibrant elements were brought together. Hungary was Ottoman, so was Baghdad. It was quite normal for there to be colorful geographical interaction. Along with other factors, the different languages and the cultural richness abounding from togetherness challenged the Ottoman citizen in every era. At the same time, the perspective of Ottoman man was immeasurably wider than that of people from other countries. In the second half of the 16th century, it was possible for a young cavalryman, if he were fortunate enough to become aid de camp, to participate in a number of campaigns, such as the Belgrade, Rhodes, Mohács, German, Corfu or Baghdad campaigns. The Ottoman man, with his army, bureaucracy and merchants stretching the limits of the various cultural forms in the different climates found on three continents, was acquainted with the world and measured it against his own cultivation. Süleymaniye in the 16th century and the Hamidiye Bosphorus Bridge, the construction of which was considered in the 19th century, are manifestations of the same comparison.

The cultural vestiges in the lands where he lived are a second factor fostering this cultivation. The demands of the Ottoman man, who was in perpetual interaction with his environment and incorporated the traces of previous civilizations into the normal flow of his life, making them his own, must have been at the highest level in all areas. Selimiye, a masterpiece of the great architect, was a response to social pressure. In the book of memoirs he had Sai Çelebi write just before his death, Sinan practically acknowledges this. On the other hand, the claim "No dome like Hagia Sophie's had ever been placed in a building under the Islamic state. It is impossible to construct one like it. We are superior to Muslims" was a veritable insult to the Ottoman man. Moreover, Sinan says in his memoirs that he found the height and diameter of the dome "excessive[7]." However, he had to say this to be politically correct. Yes, he could not surpass it[8] because he thought that a dome 55 meters high was unnecessary. Worried that it would undermine the proportionality and soundness of the building, he preferred to place upon it a dome having a height and diameter that would ensure that it would remain standing "forever." These statements in the Tezkiret'ül Bünyan (Record of Construction—the title of his memoirs) were the result of social pressure.

In the 16th century, when the State undoubtedly was the world's one-and-only pole, the importance of citizen requests must not be underestimated. The Sublime Ottoman State ruled over Arabs, Kurds, Berbers, Amalphis, Anaconas, Bulgarians, Genovese, Florentines, Pisaians, Greeks, Armenians and Jews. Nevertheless, while it was a large empire, unlike Russia, which stretched from the Urals to the Caucuses, its geographical expanse was not seamless. This engendered considerable social cost in the form of societal demands. It is assumed that this dynamic would weaken with the decline of the State over the years. Quite the opposite was true—dynamics capable of given rise to an architect-engineer the likes of Sinan remained in place. However, societal demand evolved along with the Ottoman man's experience. The expectations of the Ottoman man, who in the 16th century had expected the construction of enormous works, changed. Under the influence of architectural currents in Europe, in the 18th century, he came to demand architectural and engineering projects of a different style and dimension. The Nuruosmaniye Mosque and public works carried out in later centuries are not evidence of lapse but rather preservation. The ability to perceive and interpret this societal demand is another skill. This is just one of the important qualifications of the Ottoman ruler. The Ottoman sultan, who would wander about the marketplace in cognito, frequently sounding out the public, was a different type of ruler. This would enable him to take note of demand and respond appropriately, by either taking the measures necessary provide it momentum or to stop it in its tracks.

At the same time, the Ottoman ruler wanted to be known for his benevolence and to be the subject of prayer, which he articulated through public works. The gravestones in any Ottoman cemetery are generally higher than the walls enclosing the cemetery. The reason for this seemingly innocent practice is to stand out and come to people's attention. It is thought that the more one is noticed the more prayer will be received. There is undoubtedly a connection between this simple example and the construction of huge buildings. Anything providing the public good and facilitating the life of the subject can be understood as a request for prayer. This request can easily be observable in a variety of forms, ranging from simple inscriptions on a fountain to social complexes (külliye) consisting of huge groups of buildings. The ruler of the empire sees bringing water to places having none and building fountains, bridges, large social complexes, train stations and railways as both a public responsibility and a form of philanthropy.

He strives to maintain large public works activities and projects, which serve to reflect certain values, as well as to sustain social motivation. The appearance in İstanbul of every scientific and technological innovation made every new innovation seem banal for the Ottoman man. The age of

7 Saatçi, Suphi, *Bir Osmanlı Mucizesi,* İstanbul: Ötüken, 2005, p. 88

8 In *Tezkiret-ül Bünyan,* Sinan says "I calculated the height of Hagia Sophia's dome as 6 zira (4.54 meters) and its circumference as 4 zira (3.03 meters)." In Neslihan Sönmez's *Osmanlı Dönemi Yapı ve Malzeme Terimleri Sözlüğü* (Dictionary of Ottoman Era Construction and Materials) a Zira-i Benna is 75.8 cm. Hagia Sophia is 55.6 meters from the ground up and the diameter of its dome is between 31.24 m and 32.81. The height of Selimiye from the ground up is 43.28 meters and the diameter of its dome is 31.25 meters.

9 Rhodes, Murphey, *Süreç Siyasal Kültür Dergisi*, 1980, Issue 6, pp. 24–31

10 Bostan, İdris, "Osmanlı Bahriyesinin Modernleşmesinde Yabancı Uzmanların Rolü (1785-1819)," *İ.Ü. Edebiyat Fakültesi Tarih Dergisi*, Hakkı Dursun Yıldız Hatıra Sayısı, Issue: 35, İstanbul, 1994, pp. 177–192

11 Fendoğlu, Hasan Tahsin, *Modernleşme Bağlamında Osmanlı-Amerika İlişkileri*, İstanbul: Beyan, 2002, p. 252

12 Gülen, *op cit.*, p. 94

13 Tanrıkorur, Cinuçen, "Osmanlı Musikisi," Editor: Ekmeleddin İhsanoğlu, *Osmanlı Medeniyet Tarihi*, İstanbul, 1999, vol. 2, p. 506

14 Yasamee, F. A. K., *Ottoman Diplomacy: Abdulhamid II and the Great Powers*, 1878-1888, İstanbul,1996, p. 4

the telegraph was experienced in these lands and, shortly after its invention, the telephone became an instrument used by the Ottoman merchant. As you will read, there are dozens are examples of this policy. The driving force that mobilized this society, which was so intrepid as to take on the task of connecting the Don and Volga rivers, was the ability to make such an effort appear mundane. This proficiency had not been lost even during the Empire's so-called "sick" years; this was because it contained the arguments for its own motivation. İstanbul was a capital that still had grand dreams. The existence of the first metro on the European continent in Beyoğlu was proof that this motivation had been preserved up until the last. People who see the construction, practicality and ordinariness of Tünel, have no difficulty contemplating funiculars, sub-sea tunnels, metro lines, 700-meter-high bridges spanning the Bosphorus, huge squares and connecting the islands to each other.

During the Classical Era, Ottoman sovereigns institutionalized their desire for innovations. The Taife-i Efrenciyan (European Corps) was an institution consisting of foreign scientists and artists directly responsible to the Sultan. Well paid and working incessantly, they strove to develop good relations with the West and to relay to the palace developments they considered useful for civilian and military purposes. At the same time, in light of the information acquired, the institution was charged with recruiting the manpower needed to implement advanced technology.[9] Regardless of where they occurred in the world, technological developments were followed and either the know-how or the owner of the know-how himself was enlisted in Ottoman service.

Bringing the famous Hungarian cannonball casting master Urban, who had previously been in the service of the Byzantine Empire, into the service of (Fatih) Mehmet II, and acquiring the technical consultancy of many foreigners both before and during the siege must have been due to the efforts of this institution. Later, when overhaul of the Ottoman naval fleet was undertaken by Selim III, French, Swedish and English engineers were employed[10] The Selimiye galleon, one of the largest ships in the fleet, was built in the Golden Horn shipyard in 1795 by a French engineer.[11] Naval engineer Foster Rhodes, who was invited from the United States and worked for a long time for the state, launched a ship called Nev Eser in 1834.[12] This and other similar examples are frequently seen in the Empire's history. Not only engineers were invited. In fact, you will see later on Sultan Bayezid II (Veli) insisting on extending an invitation to Michelangelo. It was Bayezid, too, who brought Master Zeynel Adibin, who was the music director of the Horosan Sovereign, from Iran and put him into his employ.[13] As can be seen, even in the final era of the Ottoman State—in the 19th century, dozens of scientists and artists who were authorities in their fields, such as Pasteur and Bouvard, were invited, with same intent, to be guests of the Empire. In spite of everything, the Ottoman State "was at a level where they could compete technologically with its European neighbors. Plevna was never a Tel el-Kebir or Omdurman."[14]

In the same century, the peoples of the world were, in the western perspective, into two categories: "ruler" and "ruled." The "rulers" in Europe, except for one, were all Christian. There was only Yıldız Palace, which, as a Muslim empire, ruled over millions of Christians; and this situation became increasingly annoying.

This century was also the century when the concept of "empire" was questioned. The Ottoman State understood, as had Habsburg Austria, Romanov Russia and Meiji Japan, that the only way of continuing as an "empire" in the age of the nation-state was the maintenance of the "image of greatness." As important as military prowess was symbolic power capable of achieving the appearance and recognition of the state's legitimacy in the eyes of western public opinion. The way to do this was achieving recognition in the international arena of the state's legitimate political sovereignty. It was necessary to speak using symbols through which it could be understood by Europe and to be acknowledged as its equal. Not only were the solutions to this problem, which began with Sultan Mahmud II and accelerated in the Sultan Abdülhamid era with Sultan Abdülmecid and Sultan Abdülaziz Han, diverse, their intended ramifications both internally and externally were different.

Throughout his reign, Sultan Abdülhamid II reinforced the institution of the Sultan through ceremonies, public investments and philanthropy and aimed at strengthening the traditional image of the state and the Sultan in the eyes of the people. Externally, the aim, especially in 1874, was raising confidence in the Ottoman State, which had debt totaling 43 million Ottoman Liras, in international markets, to maintain the visibility of the country in the world despite the unfair images of the "Red Sultan" or the "Terrible Turk," and to counter endeavors, particularly of English politics, to undermine the legitimacy of the Ottoman caliphate.

Sultan Abdülhamid II worked at consolidating his sovereignty in every city and hamlet where people lived. However, given geographical expanse of the empire, he set about meeting his subjects, not in person but rather through a rich succession of imagery. One of the means of reinforcing control over the territories owned by the Ottoman State was investments, which represented the presence and power of the state in the area. The sultan's ubiquitous presence began to be expressed by putting the Ottoman crest on everything built in its territories. An attempt was made, especially in provincial and township centers in areas under threat of intervention, to underscore the presence of the state and make it more visible through public buildings. Inscriptions bearing the crest of the empire on state, military, civilian and religious buildings, small and large, were a symbolic representation of the sultan in the provinces, established a bond between people in towns and villages in the countryside and the sultan, and served as a means of conveying the sultan's political message. This method was also a conscious presentation of the modernization projects of Sultan Abdülhamid. By taking service to every province, despite difficult conditions, the Empire's power was not questioned by the people.

In 1875, the Ottoman budget recorded an income of 4,776,588 Ottoman Liras; this had risen to 26,015,100 Ottoman Liras by the end of the Sultan Abdülhamid II era.[15] The primary reason for this was the coupling of port towns and the countryside. The sultan undertook a substantial development plan aimed at regaining deserved Ottoman stature. First, specific commercial centers were identified in the province. Small communities were connected to these centers through highways. During the years he ruled, 16,000 km were added to the existing 900 km-highway network.[16] Subsequently, large centers were connected through railroads to one another and to port cities for easy access.

Throughout the period, there was an effort to stretch a railroad network over the entire country, which resulted in an additional 4,553 meters of new railways[17] The towns connected to one another via highways reached the ports over railways. However, the ports did not have the capacity to handle the increase in production. When the Crimean War broke out in 1853, England and France brought military and war supplies by ship to İstanbul, but because there were no docks on the coast capable of handling ships having such great tonnage, they had significant difficulties in transporting the supplies to land. In October 1895, a dock having a length of 785 meters stretching from Karaköy to Tophane was completed. Subsequently, a second enormous project to have a major impact on İstanbul's silhouette was carried out. After four years of construction, Haydarpaşa Harbor became operational in 1903. The Sirkeci, İzmir, Antalya, Beirut, Salonika, Trabzon and Mihaliç ports were rebuilt, too, because they were components of a

15 Kurthan, Şimşek, *Ankara Üniversitesi Siyasal Bilgiler Fakültesi Dergisi*, 1967, vol. 42. pp. 166–167

16 Tekeli, İ. and S. İlkin, *Cumhuriyetin Harcı-Modernitenin Altyapısı Oluşurken*, İstanbul: İstanbul Bilgi Üniversitesi, 2004, p. 157

17 Onur, Ahmet, *Türkiye Demiryolları Tarihi (1860-1953)*, Ankara, 1953, p. 32

18 Baskıcı, Murat, *Ankara Üniversitesi SBF Dergisi*, 64–1, p. 41

19 Keskinkılıç, Erdoğan, *Ankara Üniversitesi Osmanlı Tarihi Araştırma ve Uygulama Merkezi Dergisi*, Issue: 14, Ankara, 2003, pp. 319–345

20 Koloğlu, Orhan, "Yeni Haberleşme ve Ulaşım Tekniklerinin Osmanlı Toplumunu Etkileyişi," *Çağını Yakalayan Osmanlı*, İstanbul: IRCICA, 1995, pp. 604–606

21 *Sabah*, 2 Ra. 1308 (16 Oct 1890)

22 *Aksiyon*, 2010, issue 786, pp. 48–52

23 BOA, YEE, file 51, document 113, BOA, YA Hus. file 235

24 Yıldırım, Nuran, *Dünden Bugüne İstanbul Ansiklopesi*, v.5. p. 112

25 Layıktez, Celil, *Tesviye Dergisi*, issue 17, May 1995, pp. 26–27

26 BOA, YA Res. file 129, document 54

27 For examples of these practices, see Koloğlu, Orhan, *Avrupa'nın Kıskacında Abdülhamit*, İstanbul: İletişim, 2008, pp. 271–286

28 BOA, İrade Hus, 23 Ca. 1312 Genel No: 1192, Hususi No: 93

29 Ünal, Tahsin, *Türk Siyasi Tarihi*, İstanbul: Kutluğ, 1974, pp. 352–353

large economic program during this period. The sultan finally had an infrastructure through which the people could get the goods they produced to world markets. As a result of these efforts, the production of grain rose by 51%, tobacco by 191%, figs by 68%, hazelnuts by 217%, silk cocoons by 195% and cotton in the Adana region by 471%.[18] Sultan Abdülhamid lowered the Ottoman state debt, which was 43 million Ottoman Liras in 1874, to 2,640,000 Ottoman Liras in 1905.[19] Meanwhile, by 1904, an additional 26,336 km of telegraph line were added to what had been 23,380 km in 1882.[20] The country once again attained the power to handle severe waves in the open seas.

Declarations written in the name of the "Caliph of the Muslims" and delegations were sent to every region where Islam was present. To attenuate the impact of Christian missionaries in Africa and other countries, the decision was made to set up Muslim institutions;[21] Darü'l-Ulumi'l-Hamidiye University was opened in Beijing.[22] Moreover, especially in Europe, theater plays belittling Islam, Muslims and the Prophet were kept from being staged.[23] The Sultan's tolerance and scientific sensibilities were also made more visible to public opinion. The delegation chaired by Zoeros Pasha that reached Paris on 8 June 1886 gave a donation of 10,000 Francs, the largest ever given, on behalf of the empire, to Louis Pasteur for the purpose of supporting research and development.[24] Providing the most valuable assistance after the flooding disaster in the US in 1889, considered to be the "disaster of the century," was also the Ottoman Sultan.[25] In the aftermath of the Russo-Japanese War (1905), financial assistance of 10,000 Francs was sent to the parties through the Salib-i Ahmer Cemiyeti (Red Cross) to be used for health-related expenditures.[26]

Another significant practice indicative of the growth of symbolic power was being carried out by the Foreign Publications Directorate, which operated under the auspices of the Foreign Ministry. Six-hundred newspapers and magazines published in Europe were being followed on a daily basis. The writers and editors of periodicals having anti-state publications were invited to the palace, where they were personally briefed by the sultan, given various gifts, ranks and medals, provided with financial support when needed, all designed to prevent them from publishing anything opposing the Ottoman State.[27]

The idea of "Ottoman" Sultan Abdülhamid had in his head was permanently embedded in the urban fabric of Ottoman principalities through the construction of public buildings. The modernity, beauty and development of cities were photographed and photograph albums were made and sent to foreign countries. The resurrection of the empire was exhibited to the world through the architectural activities, which were the subject of these albums.

During the era of Sultan Abdülhamid, monumental public buildings were positioned at key locations on the topography of the city. Barracks, fountains, hospitals, schools, government buildings, clock towers and monuments were built on the hills outside the city walls, squares and along riverbanks. Through their dimensions, ornamentations and style, they were to be the visual and physical representatives of the power and modernity of the state. These structures were shaping urban space with new social and political meanings. The Sirkeci train station, which was built during this period, while welcoming passengers coming from Europe with an eastern ambiance, the Haydarpaşa train station showed a European face to passengers arriving from Asia. Thus, the romanticism of the Ottoman capital where East meets West was underscored.

Toward this end, throughout the 33-year reign of Sultan Abdülhamid, 72,780,000 Ottoman Liras were laid out to fulfill his command that "construction of buildings—commanding both in style and in location—befitting the fame and honor of the Ottoman government"[28] be carried out. Consequently, 1,552 national facilities were built.[29] These buildings do not include the hundreds of philanthropic works that were built through the private treasury of the Ottoman sultan, upon his personal allocation, such as Yıldız, Ertuğrul, Orhaniye, Cihangir, Adalar Hamidiye and Hidayet mosques and Şişli Etfal Hospital.[30]

As enamored by technology as Sultan Abdülmecid[31], who had had photographs of railroads hung in his chambers, Sultan Adbülhamid II was, at the end of the 19th century, an emperor who had gotten out from under debt, emerged victorious from the Greek War and had his eye set on the horizon by building the world's first torpedo launching submarine in his dockyards.[32] Farms were being mechanized, water was being brought to homes through pumping stations like those at the Hamidiye plants, dwellings and streets were being lighted by gasworks that were built, outlying areas of cities were being connected by modern boats, and new tramway lines were being set up, all of which made the Ottoman man feel as though he were once again living in an imperial city.

On account of these developments, in the first few years of Sultan Abdülhamid's sultanate, the hopelessness and pessimism that had descended upon Turkish and Muslim society had begun dissipating. The people were regaining their trust in both the Ottoman Dynasty and its governments. A new feeling of hope and expectation was forming and this expectation was turning into demand.[33]

In short, the "Sick Man" was getting back up on its feet.

It is for this reason that Bouvard was invited to İstanbul and that Arnodin, with his huge bridge project, Gavand, with is new city project, and Strom and Préault, with their projects for an undersea tunnel showed up at Yıldız Palace's doorstep. It's why Hollzman, Monsieur Moray and Karabet Efendi were in competition with one another to build a subway in İstanbul. Similarly, the milieu was such that Sarkis Balyan proposed connecting the Golden Horn to the Black Sea with a canal and Büyükada to Heybeliada (two of the Prince's Islands) with a bridge; and Osman Hamdi Bey and Ibrahim ibn lyad had the idea of building a funicular between Taksim and Kabataş. The same can be said for the development of the Suez Canal and alternative canal projects.

And dozens more incredible projects would be signed by the Bab-ı Ali (Sublime Porte) and the world would talk about this.

Turan Şahin

November 2010, Çengelköy, İstanbul

30 Aydın, Talay, *Eserleri ve Hizmetleriyle Sultan Abdülhamid*, İstanbul: Risale, 1991, p. 340

31 Karal, E. Ziya, *Osmanlı Tarihi*, VI, Ankara, 1988, p. 263

32 Gülen, Nejat, *Dünden Bugüne Bahriyemiz*, İstanbul: Kastaş A.Ş., 1988, p. 252

33 Duguid, Stephen, "The Politics of Unity: Hamidian Policy in Eastern Anatolia", *Middle Eastern Studies*, vol. 9, no. 2, May, 1973, p. 1

INCREDIBLE OTTOMAN PROJECTS

One day, while in the presence of Salih Münir Pasha, the Sultan points to a newspaper clipping he was holding and says, "Do you see this paper? This has been bothering me since yesterday." The newspaper clipping Sultan Abdülhamid had the translation of a long article on İstanbul written by a European traveler. Engaging in self-criticism, he remarks, "What he says is unfair and untrue but at the same time somewhat correct."[34]

The newspaper was holding the government responsible for not improving the neighborhoods along the coast from Sarayburnu to Yedikule. It was emphasizing that the places that will capture the attention of European travelers first, such as Eminönü Square, Karaköy Square and Galata Bridge, were being neglected. Actually, the Sultan was bothered by this situation, too. Salih Münir Pasha was an administrator who for years had lived in Europe and had a sense of art, aesthetics and urban planning. Through the full authority he acquired from the Sultan, he brought skilled engineers and masters from Europe to work to create an appearance for İstanbul that it deserved.

Sultan Abdülhamid's adoration of İstanbul had a significant impact on this capital city during his reign. He would take an interest in anything said about İstanbul and unfavorable views would upset him. This sensitivity was not the monopoly of Sultan Abdülhamid, either. On the contrary, it was a feeling that all Ottoman sultans had, to a greater or lesser extent, throughout their reigns. They wanted to acquaint themselves with the place every stone would be placed and when unable to get the information they wanted, they would intervene.

Upon the orders of the Sultan, Salih Münir Pasha contacted the French urban planner Bouvard and asked him to develop a city plan for İstanbul. Bouvard did not refuse this proposal and asked for large-format photographs of İstanbul in order to prepare a preliminary project.

As a gesture of good faith, the French government would meet the costs of the project and present it as an official gift to the Ottoman government if Bouvard was to be hired by Sultan Abdülhamid II.[35]

Bouvard's preliminary project proposed a completely new image for İstanbul. The French architect envisaged a new urban fabric of İstanbul by applying the principles of the beaux-arts in certain recognized areas of the city; his project included the removal of monuments from squares and the opening of spectacle areas. He came up with projects for the three districts in İstanbul the Sultan had selected for renovation. These were Sultanahmet Square, Beyazıt Square and the New Mosque Square. Plans for the Galata Bridge were added to these projects.[36]

34 Ergin, Osman Nuri, *İstanbul'da İmar ve İskan Hareketleri*, İstanbul: Burhaneddin Matbaası, 1938, p. 46

35 Ergin, *op cit.*, p. 47; Bouvard'ın Planları, İstanbul University Rare Works Library, No. 90591–95

36 Çelik, Zeynep, *Değişen İstanbul*, İstanbul: Tarih Vakfı Yurt Yayınları, 1996, p. 93

1 - Antonie Bouvard's Sultanahmet Square Project - 1900

Antoine Bouvard's Hippodrome Project

During the Ottoman State, the Byzantine Hippodrome was used as an open area for many different activities. Javelin contests and horse races, and such magnificent festivities as the weddings and circumcisions of the sultan's family were held in the square. What's more, it's known that (Fatih) Sultan Mehmet II did his mace training there. In a letter dated 10 April 1718 sent by Lady Montagu to Lady Bristol from İstanbul she says, "Between you and me, just as St. Paul's Cathedral in London cannot be compared to Hagia Sophie, so too is it impossible to compare any of our most well-kept squares to the Hippodrome."[37]

However, overtime, the square that had witnessed all kinds of political and cultural events of the era fell into disrepair. The temporary shanties set up by people who needed shelter in an open area in the aftermath of fires and earthquakes had transformed it into such a state that it cried out for renovation.[38]

37 Lady Montagu, *Türkiye Mektupları 1717–1718*, translated by O. Öndeş, İstanbul: Tercüman, 1978, p. 92

38 Sakaoğlu, Necdet, *Dünden Bugüne İstanbul Ansiklopedisi*, vol. 1, p. 415

The first serious step in the reorganization of the square was taken during Sultan Abdülaziz's era, with excavations of the foundations the Yılanlı Sütun (The Serpent Column) and Dikilitaş (The Obelisk of Theodosius) begun in 1856.[39] The English archeologists conducting the excavations determined the original level of the Hippodrome. In 1861, through the efforts of Naval Minister Hüsnü Pasha, the square was turned into a park for the first time.[40] Another important related initiative was taken during Sultan Abdülhamid's era. An article in *La Turquie* newspaper dated 28 October 1890 announced to its readers a project for the construction of a public park in the Hippodrome and a magnificent pavilion at either end of the square. Later on, other major renovations were made to the Hippodrome and its environs at the behest of Sultan Abdülhamid II. The Ticaret Mektebi (School of Business), Defter-i Hakani Nezareti (Ministry of Property Records), and Veterinary School were built at this time. The Sultan's second venture came just before the visit of the German emperor, Kaiser Wilhelm II in 1899. Prior to his visiting the square, it had acquired a modern appearance. The Kaiser gave a fountain to be placed in the northern corner of the square.[41]

Sultanahmet Square at the beginning of the 20th century

Bouvard's plans brought the Hippodrome to its original level. The base would reach street level by a number of symmetrically constructed steps. There was to be a monumental entrance at the northern end, which would access the Divanyolu (Road to the Imperial Council). Bouvard's 1890 plan included a previous proposal whereby the symmetric French Garden containing a pruned landscape with trees at each corner would be preserved. Dikilitaş, Yılanlı Sütun and Örme Sütun (The Walled Obelisk) provided symmetry and gave the square a Parisian, perhaps even a Concorde Square, air.

The trees to be planted along a raised walkway on two sides of the square would form an elegant border. In Bouvard's plan proposed the demolition of the Blue Mosque and Complex, which was built in 1616, along with part of its garden wall, thereby creating a sharp line perpendicular to the long façade of the Hippodrome. There was to be a typical French garden in the courtyard of the garden and the German Fountain in the middle was to be replaced by a statuesque structure with an open top.

According to the project, the 16th-century İbrahim Pasha Palace in the western part of the Hippodrome was to be demolished and in its place a police headquarters constructed. This huge building would cover the entire Hippodrome in the shape of the letter "E." To be roughly 480 m in length, it would resemble, in scale and plan, Bouvard's Industrial Palace in Paris. Plans were for new gardens along the western façade of the building, overlooking the new street to be opened and running along a north-south axis parallel to the Hippodrome.[42]

39 BOA, İrade, Dahiliye, No: 33429

40 Sakaoğlu, *op cit.*, p. 416

41 BOA, İrade, Şehremaneti, No: 1, Muharrem 1317

42 Çelik, *op cit.*, p. 93, İ. Ü. N. E. K. No: 90592

2 - Antonie Bouvard's Beyazıt Square Project - 1900

Beyazıt Square Project

While Bouvard may have wanted to preserve the historical fabric of the city in his project, he took a different approach in the design of Beyazıt Square, and submitted a proposal to make it a genuine urban center for the city. Bouvard's project called for expanding the size of the square, giving it a large rectangular shape and placing the building housing Municipal Headquarters along the axis where the Ministry of War was located. The new square would be divided into four quadrants, with fountains positioned at the intersection of the north-south, east-west axes in the center of each. The Municipal Headquarters building, which was the main component of this project, was stunning with its gigantic central tower.

The Sultan Bayezid Medrese, in the western part of the square, was to be torn down, and in its place, twin buildings with a courtyard and dome would be constructed. These two buildings were the Sanayi ve Ziraat Müzesi

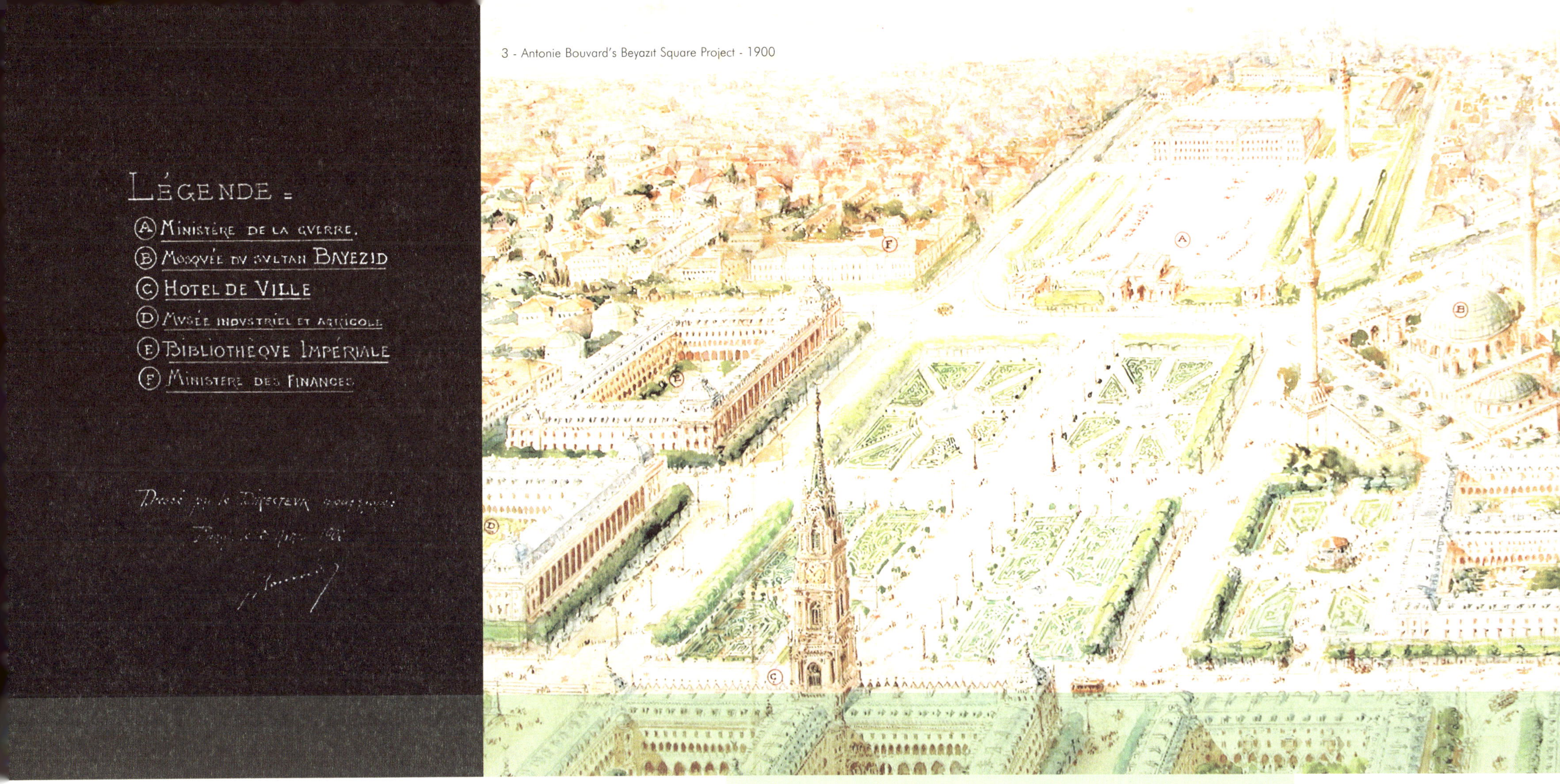

3 - Antonie Bouvard's Beyazıt Square Project - 1900

(Museum of Science and Industry) and the state library. They would be symbols modernization and progress, and education and culture.

In the southeast and southwest corners of the square, there were two similar, but unnamed, buildings. These street level arcaded buildings, with their main façades delineated by two-tiered columns and domes that call attention to the Mansard roofs and corners, have many of the main features of Bouvard buildings.[43]

43 Çelik, *op cit.*, p. 95, İ. Ü. N. E. K. No: 90593–94

View of the Blue Mosque from the Galata Bridge at the beginning of the 20th century

4 - Antonie Bouvard's Blue Mosque Square Project - 1900

The New Mosque Square Project

This project was an integral part of Bouvard's Galata Bridge idea. The southern end of the bridge had been a teaming dock since Byzantium and had heavy maritime traffic. Bouvard proposed clearing the coast and creating a large square in front of the New Mosque. The perimeter of this square would consist of two quarter circles that surrounded the mosque. Two buildings would be constructed on both sides of the mosque. In this project, the dome of the New Mosque and the new Galata Bridge would be in harmony with one another.[44]

44 Çelik, *op cit.*, p. 96, İ. Ü. N. E. K. No: 90595

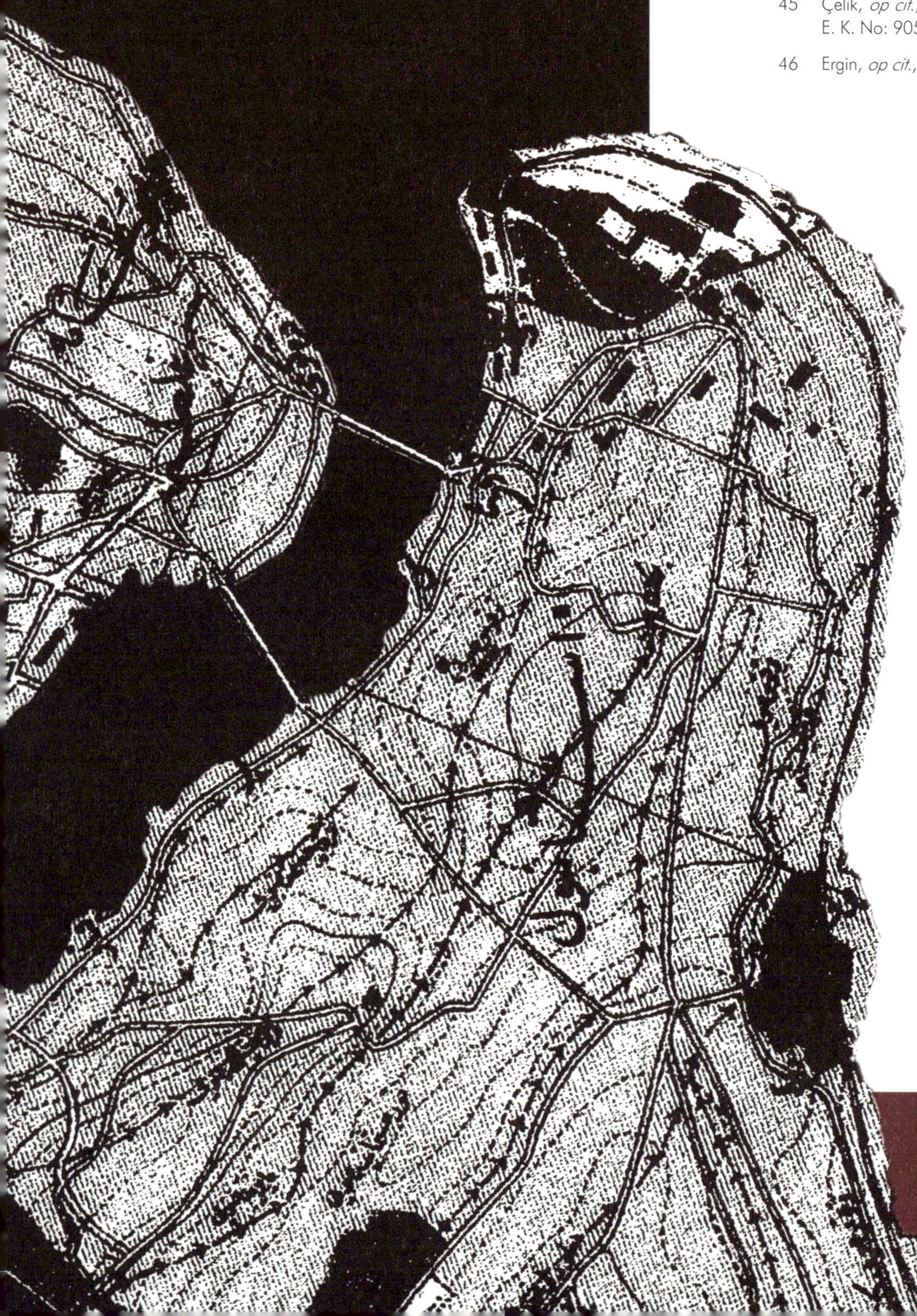

45 Çelik, *op cit.*, p. 97, İ. Ü. N. E. K. No: 90591

46 Ergin, *op cit.*, p. 48

Bouvard's Galata Bridge Project

While the former bridge's architectural design was not particularly pretentious, Bouvard was proposing a structure that any western traveler would easily recognize it as an excellent example of modern architecture. The Golden Horn in his sketch was wider than the real Golden Horn; therefore, the bridge, too, seemed longer. He emphasized the monumental dimensions of the structure of the walkways along the coast. Boulevard intended for the New Mosque, at the southern end of the prospective bridge, to serve the same purpose as Napoleon Bonaparte's mausoleum has as the focal point of the Alexandre III Bridge. The Galata Bridge, which he had designed with statues and lights on it, had two large towers at either end, thereby monumental-like entrance to the square.[45]

Despite the projects' shortcomings, they received favorable responses from the Ottoman administrators. Sultan Abdülhamid commanded his envoy to have the project implemented immediately. He said, "Tell the engineers you select that while the construction of the city's electric and gas plants, as well as the construction of electric-powered trolley lines were to be done through joint government action, we envisage awarding domestic capitalists concessions to make sure they carry out whatever planning is needed for this work." According to Osman Nuri, all hell broke loose: "By getting control of the electric and trolley business of a city like İstanbul, senior palace officials and the and shopkeeper-like minded viziers of the day, wanting to make millions through it, got their fingers in the pie, but they were unsuccessful. When laborers and foremen were later brought from France to do the job that they couldn't, the officials and viziers spread the rumor that these workers were communists and would stir up all kinds of trouble. They were able to scare the sultan sufficiently to get him to give up the idea."[46]

However, the sultan did not completely abandon this business. Looking at the photographs sent to him 6 years earlier, he asks Bouvard, who had designed the project, to come to İstanbul, which he did in 1908. This

View of the Blue Mosque from the Galata Bridge at the beginning of the 20th century

time he inspects the city, conducts analyses and then prepares a report. According to the report, which was published in *İkdam* newspaper, without there being a technical map of the city, a blueprint for reorganization is not possible. Upon Bouvard's recommendation, the job of preparing a map of İstanbul was given to the French Topography Association. An engineer by the name of Schrader was commissioned to work on it. Subsequently, the Chief Engineer of the Lyon Municipality, Aurigue, along with a number of experts, were appointed for three-year terms to the İstanbul Municipal Department of Technical Works. The project, despite much endeavor, was shelved when Sultan Abdülhamid was disposed in 1909.[47]

47 Ergin, *op cit.*, p. 49

5 - Antonie Bouvard's Galata Bridge Project - 1900

6 - A section of D'Aronco's İstanbul Agricultural and Industrial Exposition Project - 1894

7 - D'Aronco's İstanbul Agricultural and Industrial Exposition Project - 1894

D'Aronco's Dersaadet Ziraat ve Sanayi Sergi-i Umumisi (İstanbul Agricultural and Industrial Exposition) Project

Toward the end of 1892, exciting discourse and debates once again took place over an ambitious project. This time the subject on Sultan Abdülhamid's agenda was the opening of a major industrial exposition in İstanbul.

In the world of the 19th century, world fair exhibitions leading to the construction of the most noteworthy and magnificent works of architecture and engineering were organizations that both created a large-scale setting for commercial and cultural exchange and represented countries' competition for prestige.[48]

The Empire had the opportunity to participate and show off its wares at the first world fair held in London in 1851.

Believing that such an exhibition would be useful for the promotion of the country and its products, the Ottoman State did not hesitate to participate. The Ministry of Public Works and a commission set up for the purpose determined and announced which products would be sent to the exhibition, and participated jointly in the exhibition[49] Similarly, this initial experience had favorable results and care was taken to participate in subsequent world fairs.

Through the experience obtained by attending the exhibitions gave rise to the idea of opening on in İstanbul. It was hoped that by putting technical developments and the rapid and cheap production by machines, domestic businessmen would benefit in terms of industrialization.

48 Batur, Afife, "19. Yüzyıl İstanbul Mimarlığında Bir Stilistik Karşılaştırma Denemesi: A. Vallaury / R. D'Aronco", Osman Hamdi Bey ve Dönemi Sempozyumu, İstanbul: Tarih Vakfı Yurt Yayınları, 1993, pp. 146–158

49 Martal, Abdullah, *Değişim Sürecinde İzmir'de Sanayileşme, 19. Yüzyıl*, İzmir: Dokuz Eylül, 1999, p. 28

50 "Sergi-i Umum-i Osmanlı", *Dünden Bugüne İstanbul Ansiklopedisi*, vol. 6, p. 531

51 Martal, *op cit.*, pp. 30–32

52 *La Moniteur Oriental*, 13 May 1893

53 *Düstur*, 1939, Ankara, vol. 6, p. 1435

54 Batur, Afife, *İstanbul Ansiklopedisi*, İstanbul: NTV, 2010, p. 325

55 *Le Mon Arie*, 4 August 1893

56 Akçura, Gökhan, *Türkiye Sergicilik ve Fuarcılık Tarihi*, İstanbul: Tarih Vakfı Yurt Yayınları, 2009, pp. 44–45

57 National Palaces Painting Collection, Env. No. 12/2557

58 İstanbul University Rare Works Library, No. 93209

Sultan Abdülaziz got the ball rolling to get the first international exhibition in Ottoman history opened in İstanbul. The Sultan appointed the current Minister of Finance, Mustafa Fazıl Pasha, to head the preparatory committee. Following preparations, the exhibition was opened by the Sultan on 28 February 1863. The day of the opening, Sultan Abdülaziz toured the fair grounds for hours and even returned the next day. The Ottoman Exhibition received great attention from both the Ottoman public and European visitors. The exhibition continued for five months. A special, three-door building, rectangular in shape and covering an area of 3,500 m^2 was built for the exhibition in Sultanahmet Square. The building's architect was a Frenchman by the name of Bourgeois. This building was torn down two years later, in 1865.[50]

Later, the Ottoman administration continued taking part in international exhibitions by providing assistance to the domestic producer. It was announced that Ottoman citizens wanting to participate in the 1871 London and 1878 Paris Exhibitions would be exempt from customs duties.[51]

Sultan Abdülhamid signed necessary decree on 15 March 1893 to have the exhibition, which he most likely visited during his youth and observed its benefits, opened once again in the Ottoman capital. He also ordered Selim Efendi to set up a commission.

Newspapers of the day said that the fact that the undertaking coming on the Sultan's birthday was a coincidence.[52] In short order, a new by-law was prepared for the İstanbul Agricultural and Industrial Exposition, which was to open in what is today Abide-i Hürriyet Square. This exposition was considered an opportunity for the Empire to demonstrate its prowess. There was some debate over the architectural style of the buildings to be constructed there; and it was initially concluded that an "Ottoman, Arab and Maghreb," in other words, "Islamic" style would be used. Later, however, it was decided to build some of them in an Islamic, and others in a western style. An area of 142,000 m^2 was allocated for the construction of these permanent buildings.[53]

For this purpose, an agreement was made with Architect D'Aronco, who was known for having placed at various levels in competitions held at the 1887 Venice Fine Arts Exhibition, the 1890 1st Torino Architecture Exhibition and the 1891 Palmero National Exhibition. D'Aronco came to İstanbul in August 1893. He promptly prepared a presentation of his plans. Indeed, he quickly presented his watercolor plans and exhibition mock-up to the Sultan in the grand Müzayede Hall at Dolmabahçe Palace. The mock-up and drawings received the approval of the Sultan.[54] It was written in newspapers dated 4 August 1893 that D'Aronco signed a contract for this job consisting of an annual 20 thousand Francs.[55]

The exhibition would be permanent and be closed only during the winter months. Because the exhibition would be permanent, the organizing committee decided to have it built using stone, brick and iron. Visitors coming from Beyoğlu would be able to see the two facades of the main building. An area to the left of the main entrance for foreign machines and tools, along with greenhouses was proposed. The middle section was to include a large, glass-covered hall, which could be used for the opening ceremony. The area where livestock and dairy products were to be display was to be located on the right hand side of the entrance. In the northeast section, a field for agricultural experiments and right next to it a horse racing track were to be located in the northeast section. The total enclosed area, which included the Sultan's Kiosk, was 44,000 m^2. To enable easy access to all areas, a rail system was to be laid between certain spots. D'Aronco received the Legion of Merit for his blueprint.[56]

The area between Kağıthane and Darülaceze streets was being considered as the location for the construction of the exhibition. D'Aronco prepared his plans in watercolor. One the watercolor plans was for presentation and came with 12 articles noted in the lower right corner.[57] Similarly, the 29 August 1894 exhibition location plan[58] had a 14-article notation in the lower right corner. The architectural components of the plan in article form are as follows:

1- The Imperial Manor of Our Honorable Majesty

2- Great Waterfall

3- Office

4- The Grand Gate

5- Large assembly hall (Divanhane)

6- Branches of the Various Industries

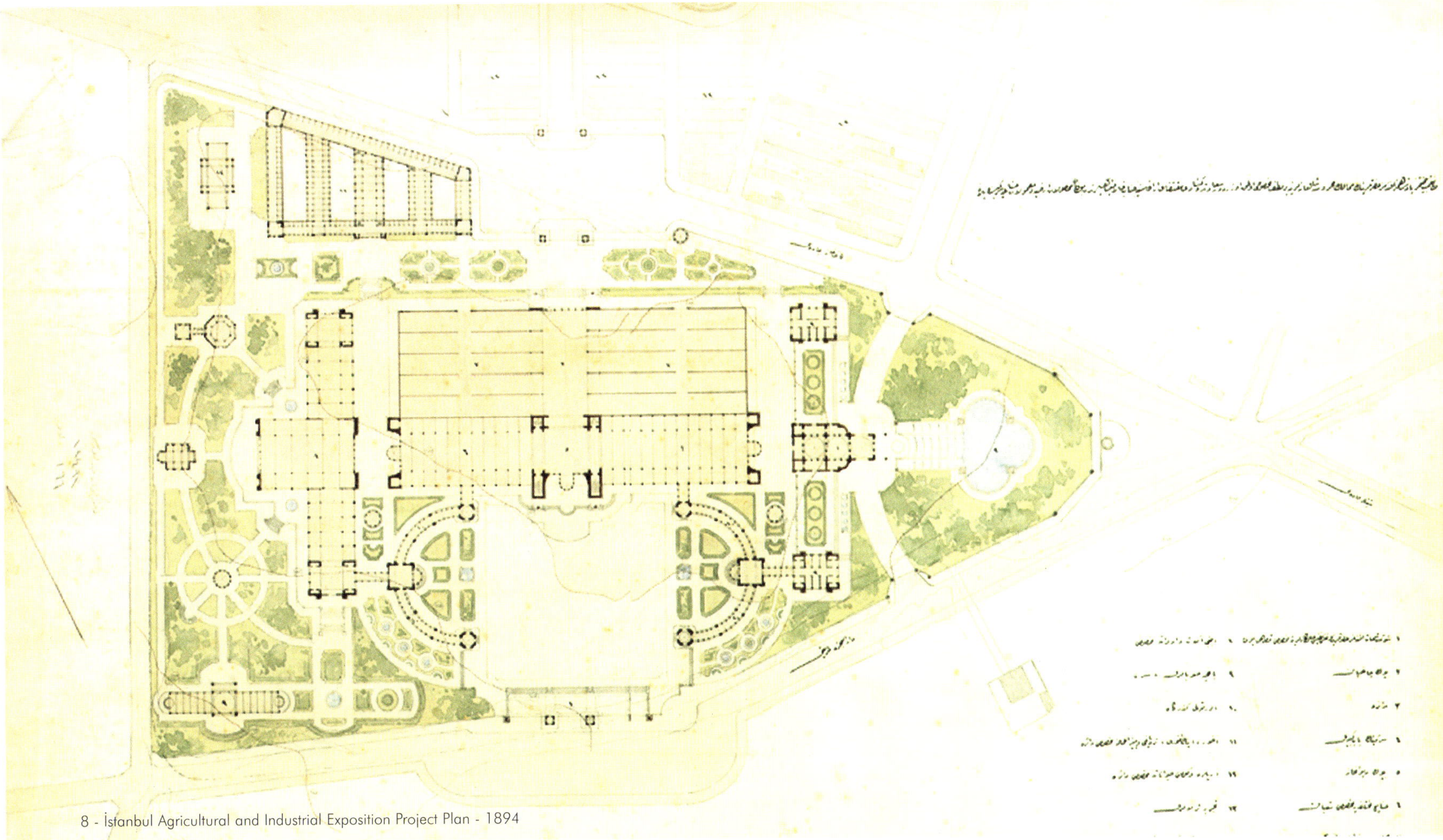
8 - İstanbul Agricultural and Industrial Exposition Project Plan - 1894

7- Mahsulat-ı Ziraiye Dairesi (Agricultural Products Office)

8- Division for foreign tools and equipment

9- Garden heaters

10-Covered Course

11-Stables, stables for cows, division for making butter and cheese

12-Division for bees and poultry

13-Practice Fields

14-Horse race track

But the 1894 Earthquake put an end to the project. D'Aronco was commissioned by the Ministry of Foundations after the earthquake for restoration work in İstanbul, for which his expertise was needed. He served the Sultan until 1909.

Three blueprints D'Aronco made for this work have been preserved. This exhibition hall, which was planned to cover an area of 142,000 m^2 in Abide-i Hürriyet Square in Şişli, was undoubtedly going to be one of the largest architectural projects in İstanbul in terms of its physical size. If we take into consideration that the current state of the Süleymaniye Complex is 70,000 m^2 and the Beylikdüzü Tüyap Exhibition Hall is 65,000 m^2, we can get a better idea of the size of the 142,000 m^2- Agricultural and Industrial Exposition Project.

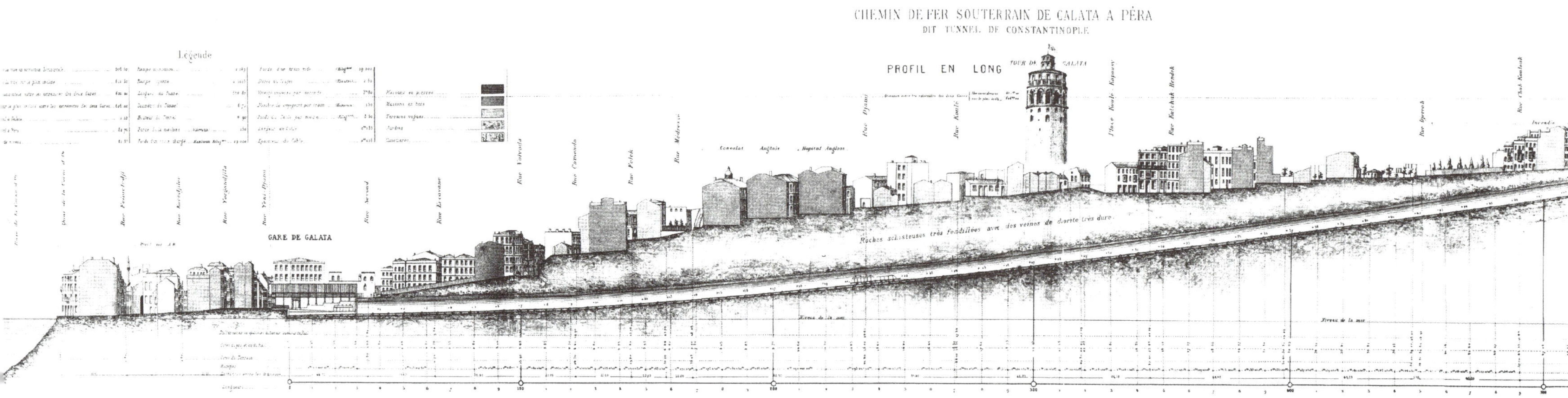

9 - Henri Gavand's Tünel Project, which went into service in 1875

Metropolitan Railway (Metro) Projects

The difficulty of building a horse-drawn tram in the centers of large and established is obvious given population density. Opening up new roads without damaging the historical fabric is a significant problem today, as well. This has given rise to the idea of laying tracks in tunnels to be dug underground and operating trains. The first place this idea was implemented was London, the capital of a former empire. Six years later, the operation of a tunnel, which was the second of its kind in the world, would begin in capital of another empire. This time, the address was İstanbul.

In May 1867, a French engineer by the name of Fuigene Henri Gavand, comes for a sightseeing trip to İstanbul. Gavand established that in those days, Beyoğlu was the most developed district in İstanbul, Galata was its commercial and banking center and an average of 40,000 people per day went up and down the Yüksekkaldırım and Galip Dede, two streets connecting them. Yüksekkaldırım was unable to handle this degree of traffic. It was six meters wide and had a 24% gradient.

Gavand had an idea that would eliminate the need for people to up and down this hill. It was possible for a subterranean passageway, working like an elevator, to carry people and goods between Galata and Beyoğlu. Calculating that the construction and operation of the project would be profitable, he takes first takes it to the French government, where it is accepted. Gavand presents this project, which now has funding following this development, to the Ottoman government on 20 July 1868. The proposal was not initially approved by the Ottoman government. The tale of the world's second metro started off full of difficulties. In brief, the agreement was signed on 3 November 1869, but this time a war broke out between France and

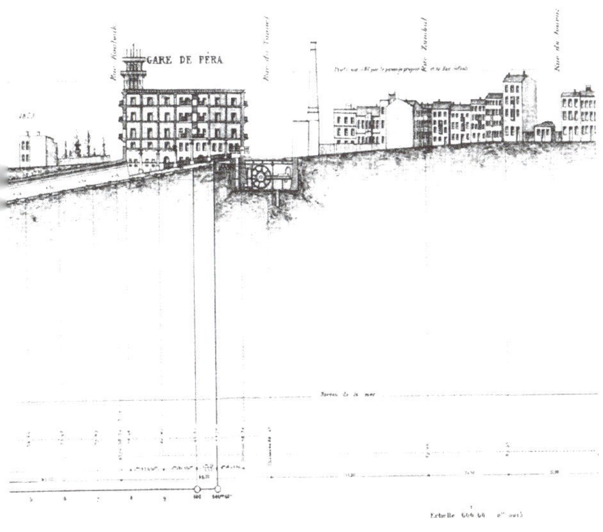

Germany in 1870, which led to France pulling out of the project. Gavand works as if the agreement was not annulled so that he could gain time to find a new investor. This situation did not escape the attention of Ottoman administrators. The project would remain suspended until English support was found in May 1871. After this date, work picks up and the world's second metro becomes operational in January 1875.

The tunnel (Tünel) is 554 m long, 6.70 m wide and 4.90 m high and consists of a station and machine room at either end. Two, two-wagon cars operate on a single line. One of the wagons is allocated to passengers and is separated into stations. At each station, men and women sat in separate compartments. The second wagon was for goods, animals and the transfer of horse-drawn cars.[59]

Undoubtedly, the metro efforts during the Sultan Abdülhamid era had become commonplace for both the Ottoman subjects and administrator.

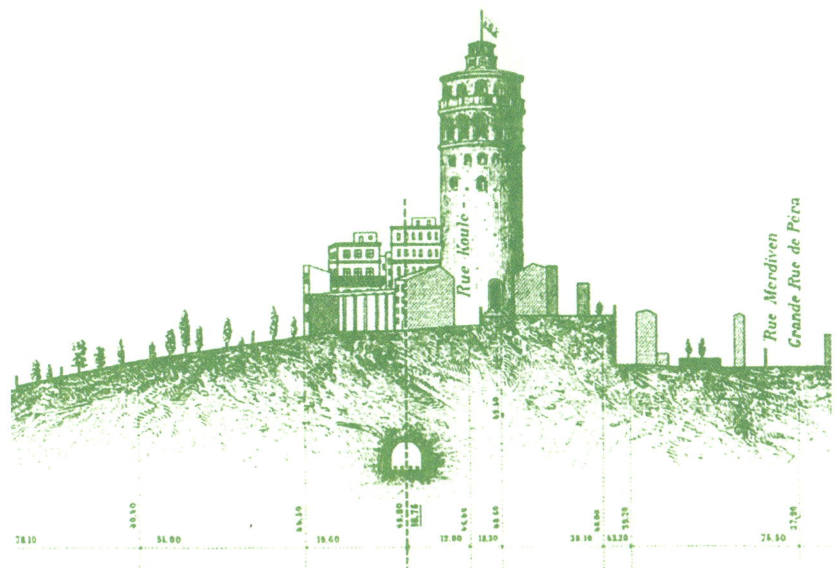

59 Engin, Vahdettin, *Tünel'den Feniküler'e*, İstanbul: Yapı Merkezi, 2007, pp. 33–55

10- Henri Gavand's Yeni Şehir (New City) Project - 1876

Once Tünel went into service in 1875, even much bolder projects were proposed to the palace. The first proposal, once again, came from Gavand.

Toward the end of 1876, Gavand, the ingenious entrepreneur in Ottoman lands who got Tünel built, presented a project to the brand-new Sultan to build a metro, a major part of which would be underground, in İstanbul that would run from north to south; a large port that would run along the coast between Yedikule and Sarayburnu; and the establishment of a new city on the filled in area. Gavand called this project "Yeni Şehir" (New City).[60]

Gavand is known to have developed many projects for İstanbul during the four-year construction of the tunnel; he also had analyses conducted and made proposals to eliminate the shortage of water in the city.

In the presentation of his proposal, Gavand summarized why such a project was needed and offered salient findings: "The Commercial and Military Port, located on the Golden Horn, was quite small given the conditions at that time. The Ottoman fleet had grown to a remarkable degree. Today, Turkey is the third or fourth largest naval power in the world. The number of ships coming to İstanbul from all around the world is constantly increasing. The physical structure of the Golden Horn and the Bosphorus prevent the expansion of existing harbors. It's also impossible to construct the facilities for loading, unloading and storing here, because homes and shops have come right now to the coast. The cost of taking them for public use will be expensive. For a long time, given this impracticability, the idea of building another port in İstanbul has been on the agenda. The area between Yedikule and Sarayburnu is a natural bay, and provided that it is filled in, it could be used to build a new city. The advantage this gigantic endeavor would provide to the government and commerce will be extraordinary. The government could also use the Golden Horn as a whole as a military harbor. A central customs office that would double major courses of revenue could be set up. As the result of the formation of institutions being considered for Yenişehir it will be possible to reach the level of prosperity achieved by leading English cities. Certainly, this new city, harbor and warehouses will make İstanbul one of the most important maritime cities in the world; and it will have commercial importance that parallels its political significance."[61]

Gavand conceived of Yenişehir as being connected to different piers and railways; therefore, he also planned Metro and undersea tunnel passages. The railway, which would begin in Kumkapı, on the coast of the Sea of Marmara, would go underground through the peninsula, and exit at Eminönü, where a terminal was to be built. Vehicles would use the new Galata Bridge to go over the Golden Horn and then go underground again. According to the project, the system would run parallel to the coast from Karaköy to Ortaköy, and there would be stops at Karaköy, Tophane, Fındıklı, Dolmabahçe, Beşiktaş and Ortaköy. The tunnels in Gavand's plan totaled 4,300 m in length. The first tunnel was planned to be 1500 m, the second tunnel 600 m, the third tunnel 700 m, the fourth tunnel 800 m and the fifth tunnel 700 m long.

Gavand failed to provide details concerning the tunnels to be constructed for the metro that was to extend from Beşiktaş to Kilyos on the Black Sea coast. His plans consisted, though, of much more than this. It was thought that Yenişehir needed to be linked to the Anatolian side of the city. Therefore, his idea was for Sarayburnu and Üsküdar to be connected by an underwater train running below and across the Bosphorus. But the particulars of this ambitious project were not included in a proposal. Gavand's intercontinental metro project would be the source of inspiration for many undersea tunnel projects presented to the Sultan in the years to come.

The harbor project prepared by Gavand called for the filling in of part of the coast between Yedikule and Sarayburnu. The plan called for the construction of a large harbor and a new city around this area. The project, called the "Yeni Şehir" (New City) Project, which was submitted to the Ottoman government, included blueprints for a metro and called for the clearing of 2,200,000 m^2- parcel of land. The earth obtained from the excavation of the terrain would be used to fill in the sea in front of what would be the harbor. In the end, a harbor covering an area of 2 million m^2 would be created. It would be 3,760 m long and 600 m at its widest point. Gavand did not include the cost of the metro in his cost estimates; he only what it would cost to build the harbor and "Yeni Şehir:" "An expenditure of at least 300 million Francs is needed to carry out this extensive project. However, expenditures will be commensurate with revenues. For example, after filling in the sea and building the dock, port and walkway, there will

60 Erinç, Orhan, *92 yıl önce İstanbul Metrosu Çalışmaları ve Yeni Liman Projesi*, BTTD, 1968, issue 7, pp. 48–51

61 *İstanbul Tüneli*, translated by Vahdettin Engin, İETT, 2011, pp. 50–52, Gavand Eugene Henri, *Tunnel de Constantinople*, Paris, 1876 (One copy is in the Archeological Museum and the other is in the Çelik Gülersoy Library in İstanbul).

be a 1 million 500 thousand m^2-track of land remaining. When this land, valued at an average 300 Franc per square meter, is sold, a return of 450 million Francs can be made."

Fees would be charged ships entering the port and a docking tax would be levied on ships docking there. The revenue from docking fees would be enormous. Even so, Gavand admitted that while revenues would be good, given the significant cost of the project, they may not be sufficient to meet expenses. At any case, he argued that the metro had a better chance of becoming a reality since dignitaries lived on the Bosphorus and a number of political advantages would be forthcoming. The ramifications of the Mithat Pasha Government fiasco of going to war with Russia in 1877 were to be great, resulting in the suspension of this and similar projects for a while. Despite the repercussions of war, because of the resurgence of imperial dynamics, this interlude was shorter than previously anticipated. With economic recovery, investments once again began to be made. Equity capital and proposals of both domestic and foreign entrepreneurs would be taken into consideration. One of these entrepreneurs was Bogos Efendi, whose proposal was similar to Gavand's project.

Archive records show that Bogos Efendi's proposal came 1887—11 years after the war. Bogos Efendi was a civil servant at the Reji İdaresi.* He was an Ottoman citizen. With his petition dated 27 September 1887, Bogos Efendi requested a 99-year franchise to construct a metro running from Bahçekapı to Beyazıt. He also declared that he was prepared to furnish the capital

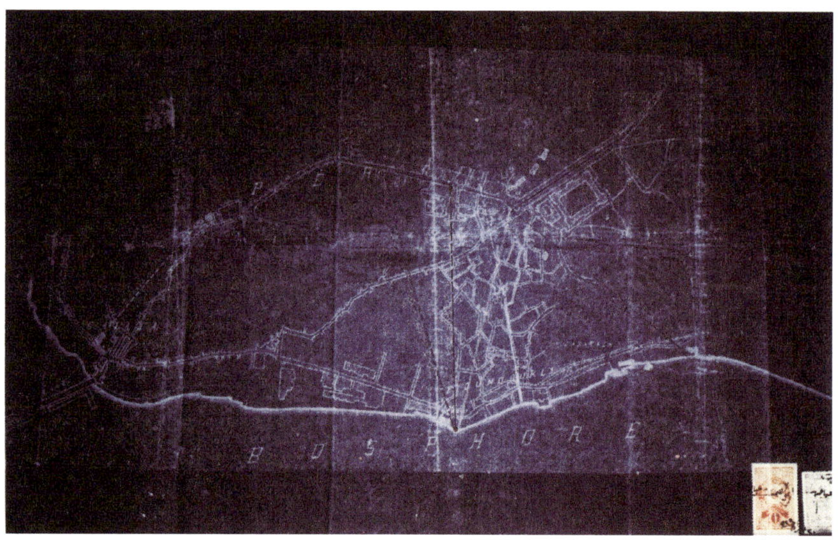

11 - The railway line from Salıpazarı to Beyoğlu - 1908

necessary for this business, that he would create an Ottoman incorporated company subject to Ottoman laws, and if a franchise was granted, that he would deposit 2,000 Liras in a bank chosen by the government as assurance that he would meet his obligation.[62] Bogos Efendi's was deemed unnecessary and rejected by the Council of Ministers.[63]

The idea of constructing a metro between Bahçekapı and Beyazıt was to be reexamined 10 years later. This time, the person coming up with the scheme was Namık Pashazade Tahir Bey. In March 1898, Tahir Bey asked to be granted a franchise for the construction of a metro running along a Bahçekapı-Çarşıkapı-Beyazıt route. Tahir Bey's request would be considered within the framework of applications from others who had petitioned to build the same line. The rejection of Bogos Efendi's petition was taken as a precedent. The proposal was rejected on the grounds that it would cause property prices to drop and be detrimental to merchants and the state.[64]

Without a doubt, this matter would continue to be pursued. The work of Osman Hamdi Bey, an archeologist, museum curator, painter and teacher at the Sanay-i Nefise Mektebi (School of Fine Arts) caught the attention of Sultan Abdülhamid. Osman Hamdi was the director of the Sultan's Müze-i Hümayun (Imperial Museum). He had thought about the country's problems and had come up with solutions for them.

A tunnel had been offered as a solution to the heavy pedestrian traffic on the steep hill between Yüksekkaldırım and Galip Dede streets; this gave Osman Hamdi the idea of doing the same between Taksim and Kabataş.

Hamdi Bey shared his project with the government in 1895.[65] The government passed the project on to the Ministry of Commerce. In his petition, Osman Hamdi Bey proposed a system of funiculars run on narrow tracks and operated by steam engine. What's more, in the supplement of the *Servet-i Fünun*** dated 4 July 1895, it was said that "In the Domestic News Headline consent was given to Osman Hamdi, Director of the Müze-i Hümayun, to build a funicular railway line from Kabataş to Taksim."[66]

Previous metro proposals had envisaged service on the European side. The first to recognize this deficiency was Monsieur Kirbis. The justifications for building a funicular line between Karaköy and Galata were equally valid for constructing one between Üsküdar and Bağlarbaşı. Many people were

* Tr. note: A monopoly set up by the French to process tobacco in the Ottoman Empire in exchange for debt payments.

62 BOA, BEO, Nafia Giden, no 81733

63 BOA, MVN, no 27, Provision 4

64 BOA, BEO, Nafia Giden, no. 81733

65 Ender, Celil, "Osman Hamdi Bey'in İmtiyaz İstediği Kabataş-Taksim Metro Projesi", Osman Hamdi Bey ve Dönemi Sempozyumu. İstanbul: Tarih Vakfı, 1993, pp. 172–175

** Tr. note: A leading Turkish literary magazine

66 *Servet-i Fünun*, 4 July 1895, Issue. 225, Supplement November, p. 5

trekking up the steep hill from Üsküdar Square to Bağlarbaşı. The state of affairs was ripe for social and commercial initiatives. Monsieur Kirbis, declaring that he worked on behalf of Kay Factories, gave Tünel as an example in his petition. He asked for a franchise "to carry passengers up and down the hill in wagons pulled by a cable operated by a stationary machine, just like the system used underground between Galata and Beyoğlu." The commission set up to evaluate the petition submitted by Kirbis Efendi, who declared that, if necessary, he would set up an Ottoman Incorporated Company, issued a favorable review of said petition. The date of this initiative was 6 August 1894.[67]

However, for investors, the European side had not yet lost its attraction as a center. 13 years after Osman Hamdi Bey, in a petition given to the Ministries of Public Works and Commerce on 25 December 1908, it was proposed that railway line from Salıpazarı to Beyoğlu, pulled by cables and operating by steam, be established. The institution making the application was the company Agence Generale Panhard & Levassor, located on 21 Mumhane street in İstanbul. It applied under the name Müessesat-ı Memalik-i Osmaniye Umum Acentası (Public Agency of the Institutes of Ottoman Countries) Abdullah ibn Iyad and Partners. They requested a 50-year franchise in the event the projects were approved. The company to be set up to build the railway line was called the "Salıpazarı-Beyoğlu Osmanlı Funicular Company."[68]

When Tünel Company, which was engaged in trying to get a concession for the construction of a line that would go from Salıpazarı to Beyoğlu, heard about this other proposal, it immediately took action; it felt the need to remind those involved of Article 30 of the Terms of Contract concluded on 6 November 1869. According to this article, Tünel Company had preemptive right concerning railway lines to be built around Tünel. Accordingly, in the deliberations conducted, the company objected to the Salıpazarı-Beyoğlu line. Taking into consideration the difficulties the company would encounter in the future, it opposed the proposal made by Abdullah ibn Iyad and his partners. Nevertheless, the company that had submitted the petition was insistent; first, it replaced its partners and then it changed the route of the funicular line. The new route was from Taksim-Fındıklı, but this time the existence of another obstacle made the petition futile. Article 27 of the contract made with the Beyazıt Metropolitan Railway

12 - Osman Hamdi Bey's Bağlarbaşı - Üsküdar Funicular Project - 1895

67 BOA, Y. Mtv, no 102/18

68 BOA, T. NFİ, no 1364/22, lef 3–4

13 - Lacey Sillar - Leigh Westminster's Şişli - Beyazıt Metro Proposal - 1909

Company, which had assumed responsibility for the construction of the metro between Beyazıt and Şişli constituted an impediment to the building of a metro between Fındıklı and Taksim.[69] Application was made many times afterwards. Because there was a need for this line, it was discussed by many companies. The final deliberation occurred in February 1915.[70] As for the Beyazıt Şişli Metropolitan Railway Company, it had laid on the table Article 27 of the contract it had made with the Ottoman Government. At the conclusion of intense negotiations, it was able to get the project approved, on the basis of national interest, as defined by the recently reorganized Ottoman bureaucracy. An edict granting a franchise was issued on 15 October 1912 and the agreement was signed on 26 January 1913.[71]

According to the agreement, the Sublime Porte "confers upon Monsieur Edouard Huguenin the right to construct and operate a city railway, referred to as "metropolitan," for the transport of passengers and commercial goods on the Rumeli side of İstanbul and its environs. "The terms of the agreement[72] calls for a main line to be built initially between Beyazıt and Şişli. The three stops on the main line would be connected to one another by auxiliary lines; the project would thus encompass İstanbul from one end to the other. The central point of the auxiliary lines was to be the Beyazıt Station. One line was to go from Beyazıt to Yenikapı and another to Eyüp; a line running from Eyüp to Dolmabahçe would also be built. The colossal project, which aimed at connecting the outermost parts of the city through a network of stops such as Dolmabahçe, Şişli, Eyüp and Yenikapı, unfortunately, would fall victim in later years to the political upheaval of the period.

When the agreement was signed, Gazi Ahmed Muhtar Pasha was the Grand Vizier, but two weeks later, on 29 October 1912, he was forced to step down. Kamil Pasha was brought to the position vacated by Muhtar Pasha. Meanwhile, the Balkan Wars were continuing. On 23 January 1913, the Sublime Porte was stormed and Mahmud Şevket Pasha was made Grand Vizier. This political upheaval, the Balkan Wars and the uncertainty that followed meant that one of the largest public works and projects in Ottoman history would remain inconclusive.

The contractor partner in this project of Monsieur Edouard Huguenin, who had, in partnership with Ottoman Bank and many other investors, and had acted on behalf of Deutsche Bank, was Phillip Hozman A.G., contractor for large projects in Ottoman lands during the rule of Sultan Abdülhamid II.

Phillip Hozman A. G. was a small family company founded near Frankfurt in 1849. In time, it grew and, with the support of Wilhelm II, made railroad investments in dozens of countries, including the Ottoman State. In fact, it was one of the companies initially involved in the Baghdad Railroad project, which, although having gotten off the ground in 1888, was not completed until 1940. It was responsible for putting up impressive buildings that altered the landscape of İstanbul, one of which was the Haydarpaşa Train Station. Philipp Holzmann A.G. also built 1,024 km of railroad tracks in the country. Working with partners, the company completed the Haydarpaşa-İzmit line in 1888, the İzmit-Arifiye line in 1890, the Arifiye-Eskişehir line the same year, the Eskişehir-Ankara line two years later, the Eskişehir-Kütahya line in 1894, the Alayurt-Afyon line in 1895 and the Afyon-Konya line in August 1896. It also built the stations and dozens of bridges on these lines.[73]

İstanbul had lost its chance of hosting a huge project with a major investor, construction company and big financiers like the Ottoman Bank and Deutsche Bank, but there were still many who were willing to push the limits. This time, it was Monsieur Moray, the project proposer, and Lacey Sillar and Leigh Westminister, project engineers.[74]

Monsieur Moray asked for a franchise to build a metro from Taksim to Galata, from Galata to Sirkeci, and from there to Beyazıt. He made the request, but despite insistence on the part of the government, he failed to prove his financial worthiness. The Ottoman bureaucracy appears to have acted fastidiously when it came to such matters. However, a bad example had been set for other potential applicants by discussing franchise terms with him and preparing an agreement without first having received this evidence from him. Therefore, before negotiations could be conducted with Moray, he had to demonstrate that he had sufficient financial means to carry out such an endeavor. When Monsieur Moray was unable to provide such a guarantee, his proposal was left hanging in the air.[75] It can be said that political instability played a major role in Moray being unable to attract investors in his proposal.

69 BOA, ŞD, no 510/1 lef 19

70 Engin, *op cit.*, pp. 175–183

71 BOA, Meclis-i Mahsus İradeleri, no 36, 4 Za 1330; BOA, MVM, no 227, Provision 244

72 For the contract, see Osman Nuri, *Mecelle-i Umur-u Belediye*, v. III, pp. 293–321

73 Pohl, Manfred, Philipp Holzmannn: *Geschichte eines Bauunternehmens, 1849–1999*, München: C.H. Beck, 1999, pp. 100–107

74 İBB Atatürk Kitaplığı Hrt. 006226 No. Catalog

75 BOA, BEO, Nafia Giden, no. 269028

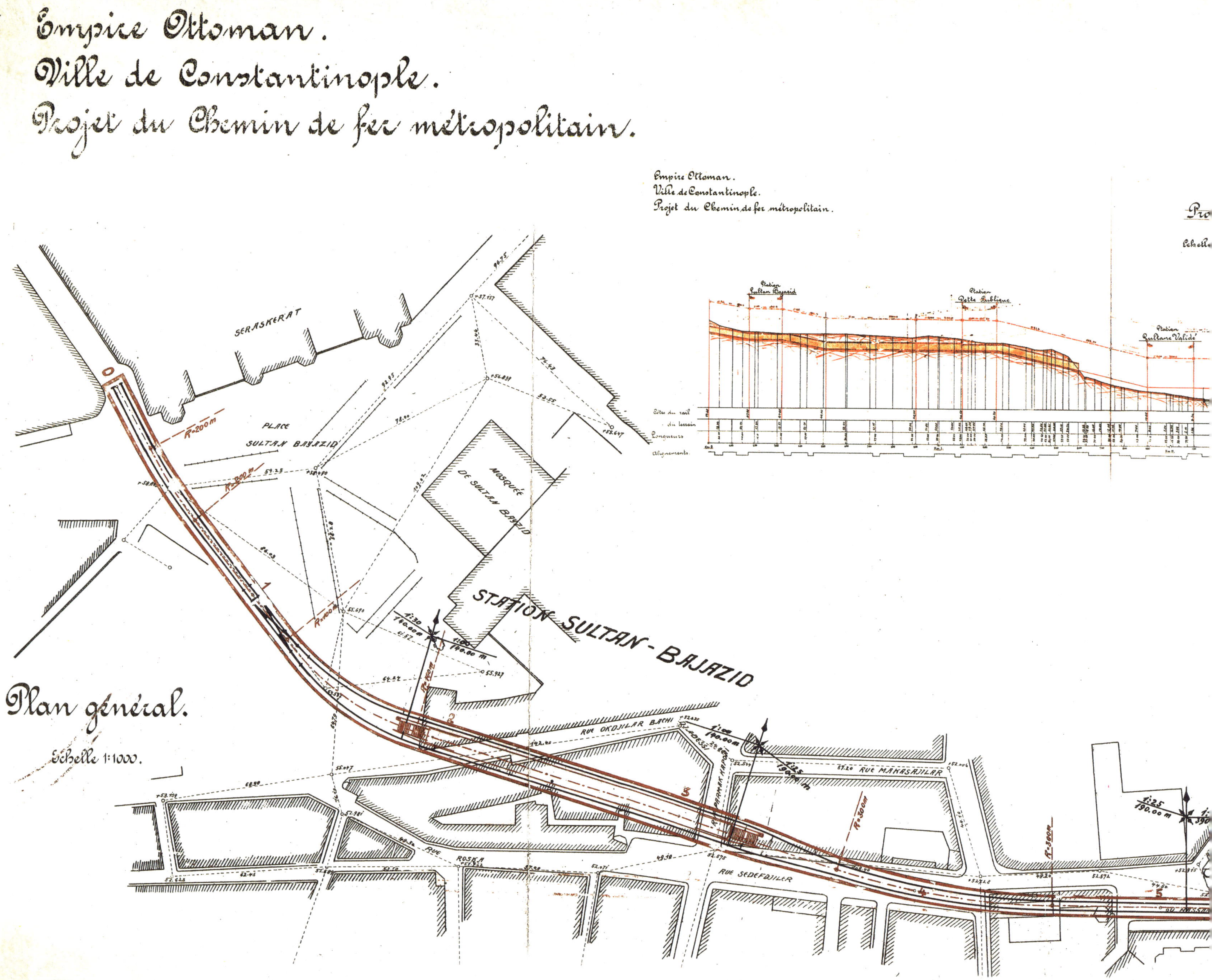
Empire Ottoman.
Ville de Constantinople.
Projet du Chemin de fer métropolitain.
Empire Ottoman.
Ville de Constantinople.
Projet du Chemin de fer métropolitain.
Station Sultan Bajazid
Station Dette Publique
Station Sultane Validé
SERASKERAT
PLACE SULTAN BAJAZID
MOSQUÉE DE SULTAN BAJAZID
STATION SULTAN-BAJAZID
Plan général.
Echelle 1:1000.
RUE OKDJILAR BACHI
RUE MAKASJILAR
RUE ROSKA
RUE SEDEFDJILAR

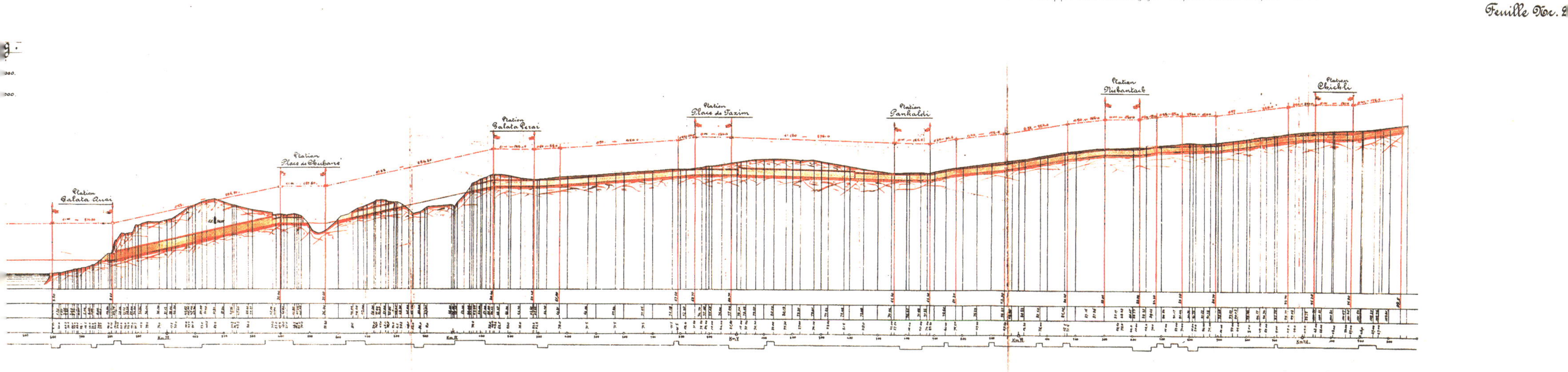
Feuille Nr. 2
Station Galata Quai
Station Galata Serai
Station Place de Taxim
Station Pankaldi
Station Nichantach
Station Chichli

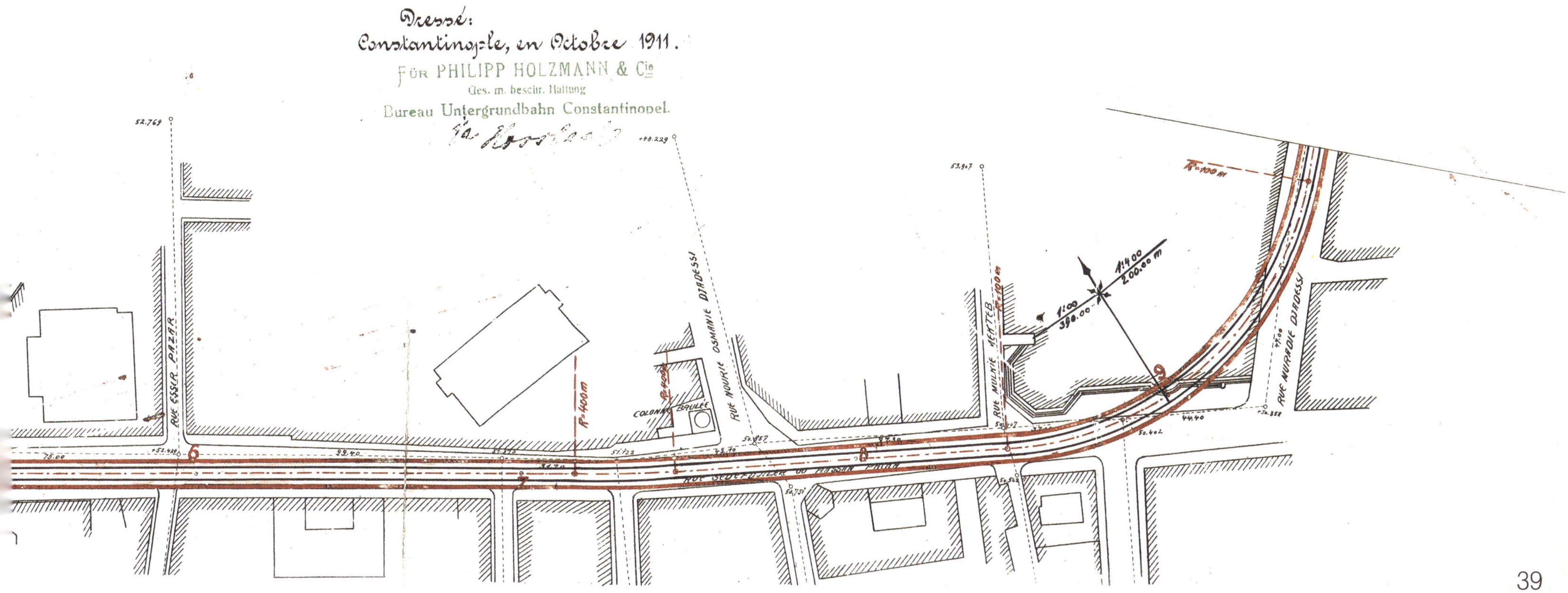
Dressé:
Constantinople, en Octobre 1911.
Für PHILIPP HOLZMANN & Cie
Ges. m. beschr. Haftung
Bureau Untergrundbahn Constantinopel.
RUE ESSIR PAZAR
R=400m
COLONNE BRULÉE
RUE NOURIE OSMANIE DJADESSI
RUE MOUHIE MEKTEB
R=100m
1:400
200.00 m
RUE NURADIE DJADESSI

Empire Ottoman.
Ville de Constantinople.
Projet du Chemin de fer métropolitain

a. Coupe transversale pour les lignes souterraines

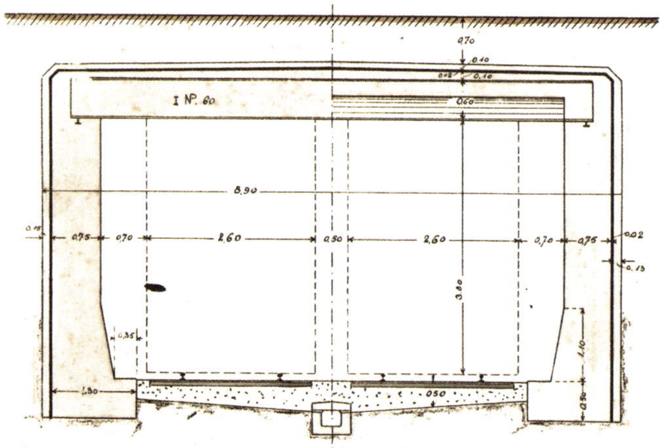

b. Coupe transversale du profil avec appuis intermédiaires.

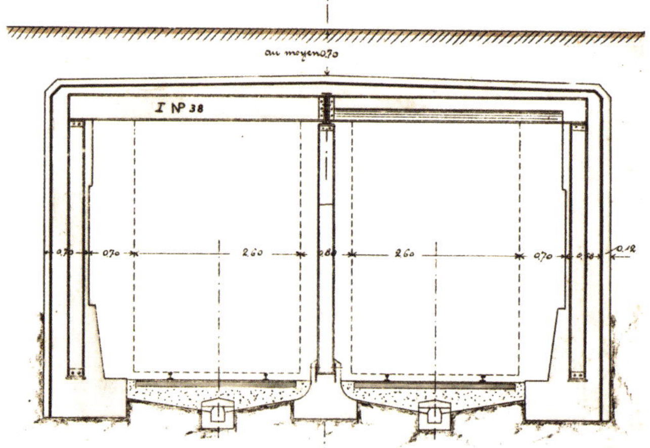

c. Coupe longitudinale Profil b

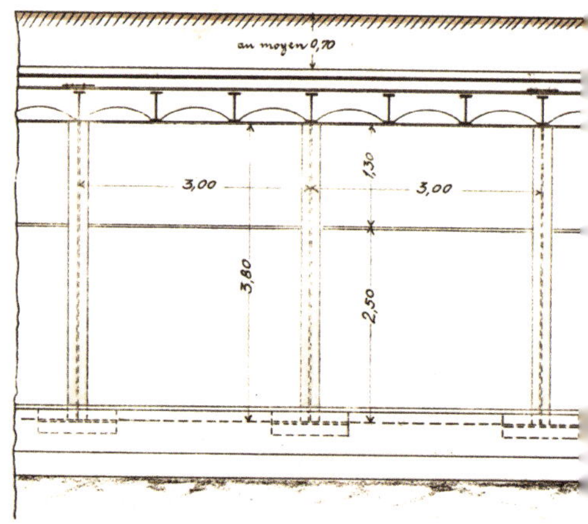

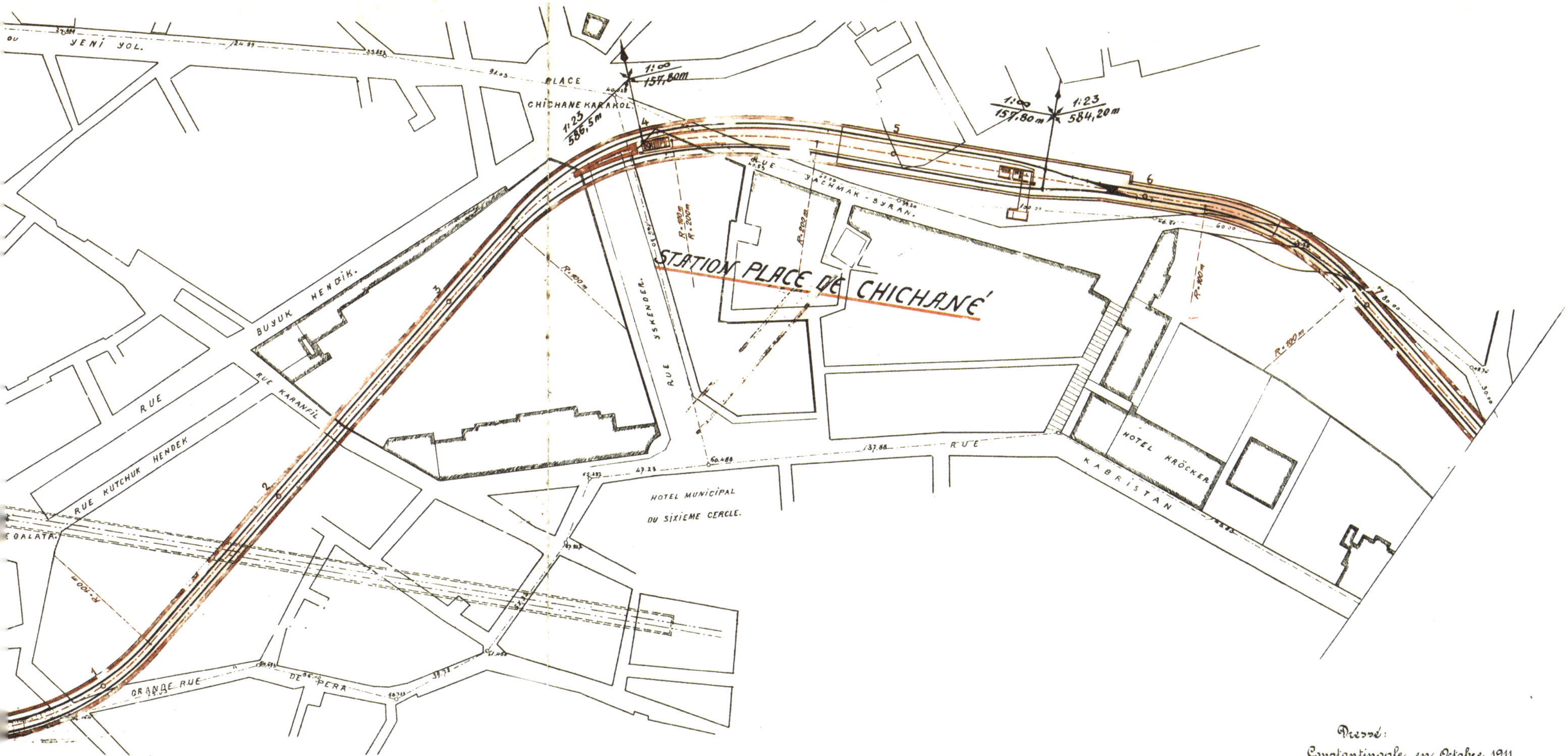

Dressé:
Constantinople, en Octobre 1911
Für PHILIPP HOLZMANN & Co
Bureau Untergrundbahn Constantinopel

ses coupes
lle 1:50.

d.
Coupe transversale des tunnels
à simple voie.

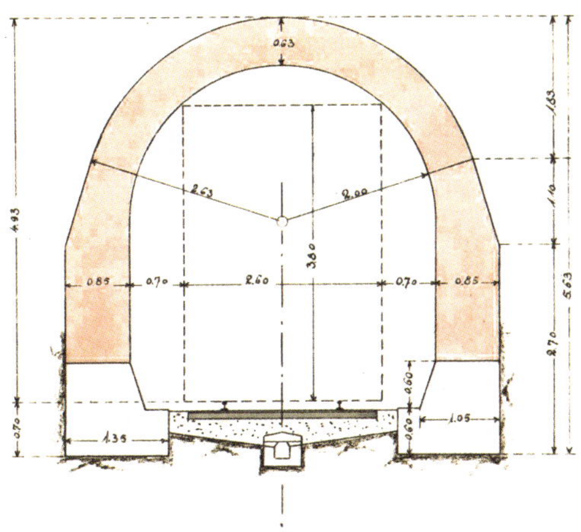

e.
Coupe transversale des tunnels
à double voie.

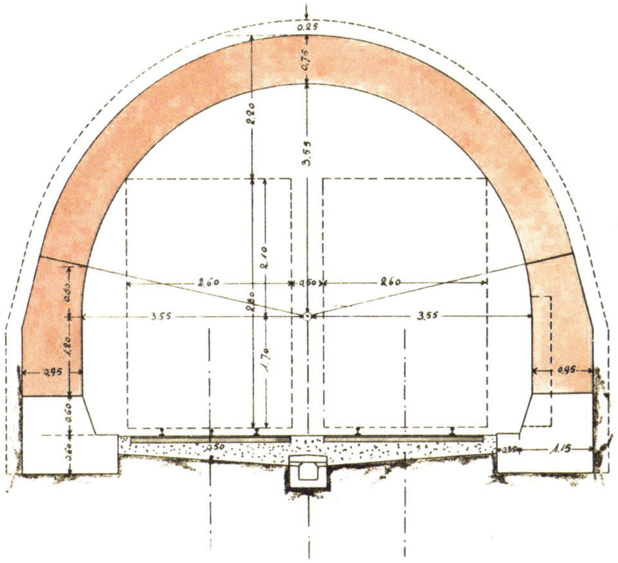

f.
Coupe transversale
sur les embranchements

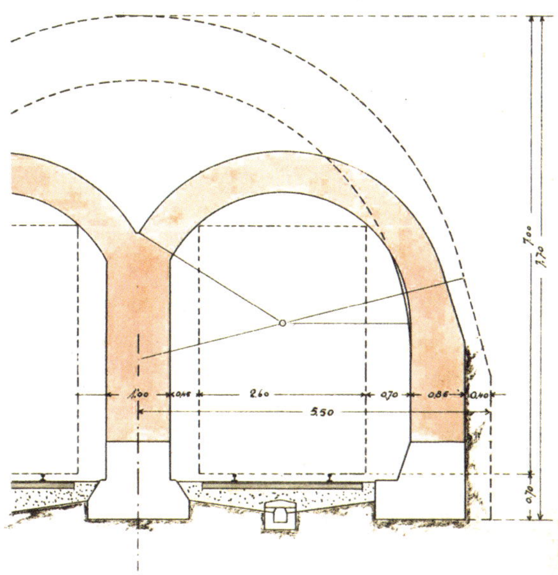

Sections from Philipp Holzmann's Şişli - Beyazıt Metro Project

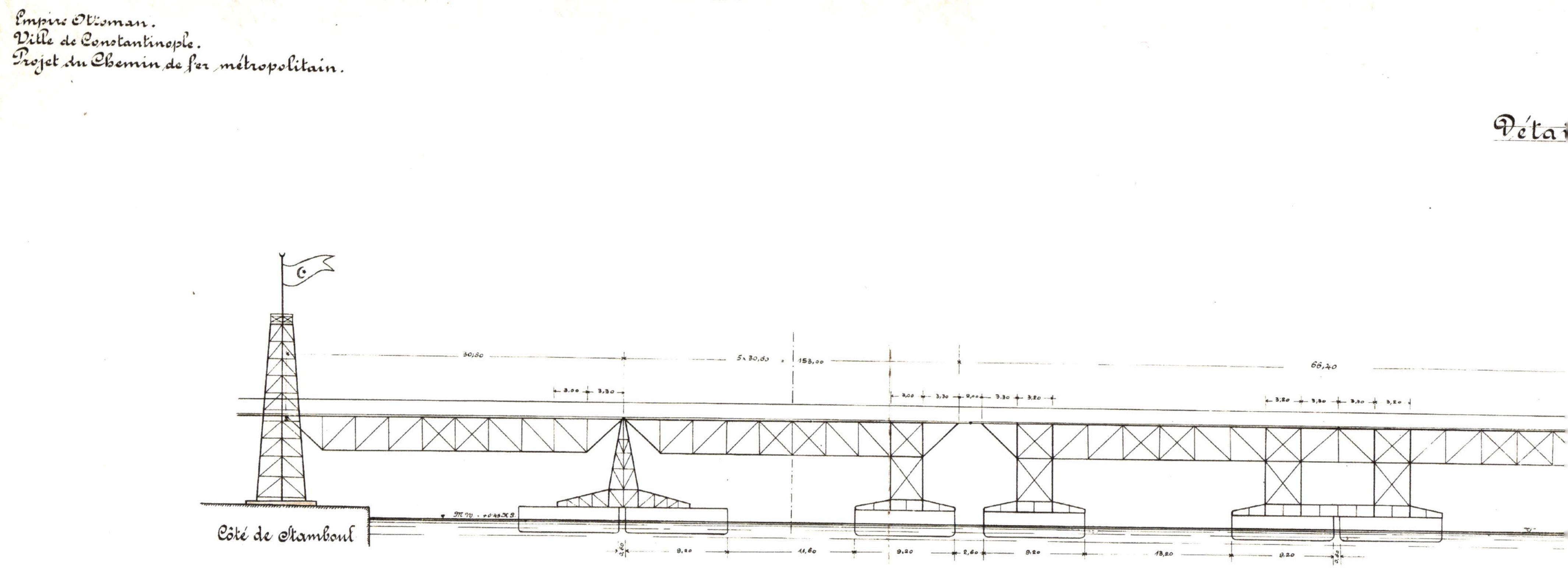
Empire Ottoman.
Ville de Constantinople.
Projet du Chemin de fer métropolitain.
Détai
Côté de Stamboul

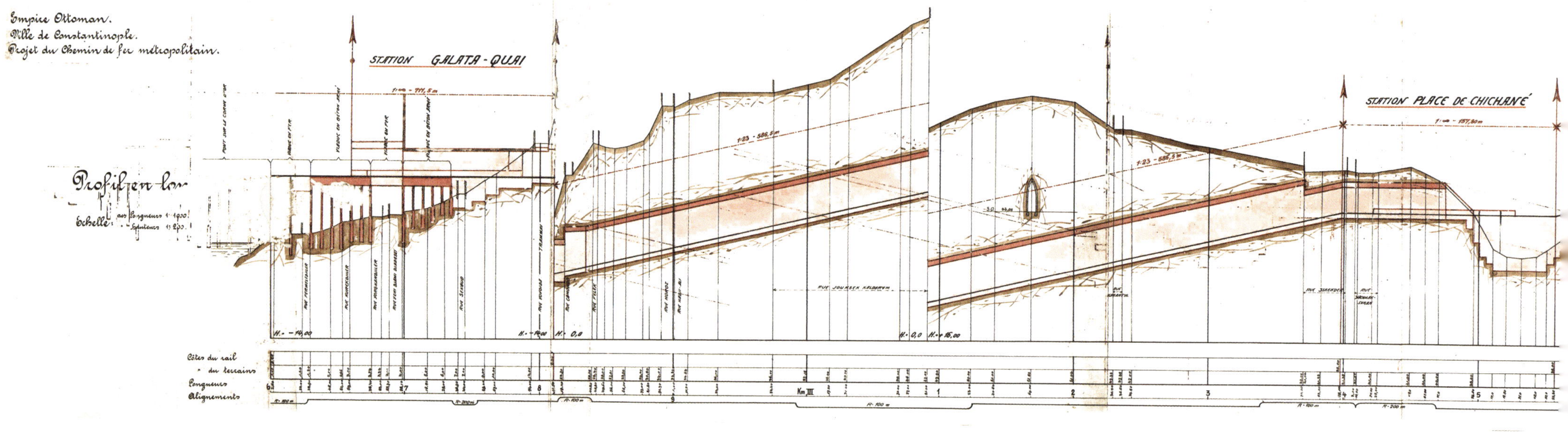
Empire Ottoman.
Ville de Constantinople.
Projet du Chemin de fer métropolitain.
STATION GALATA-QUAI
STATION PLACE DE CHICHANÉ
Profil en lon
Côtes du rail
du terrains
Longueurs
Alignements
Km III

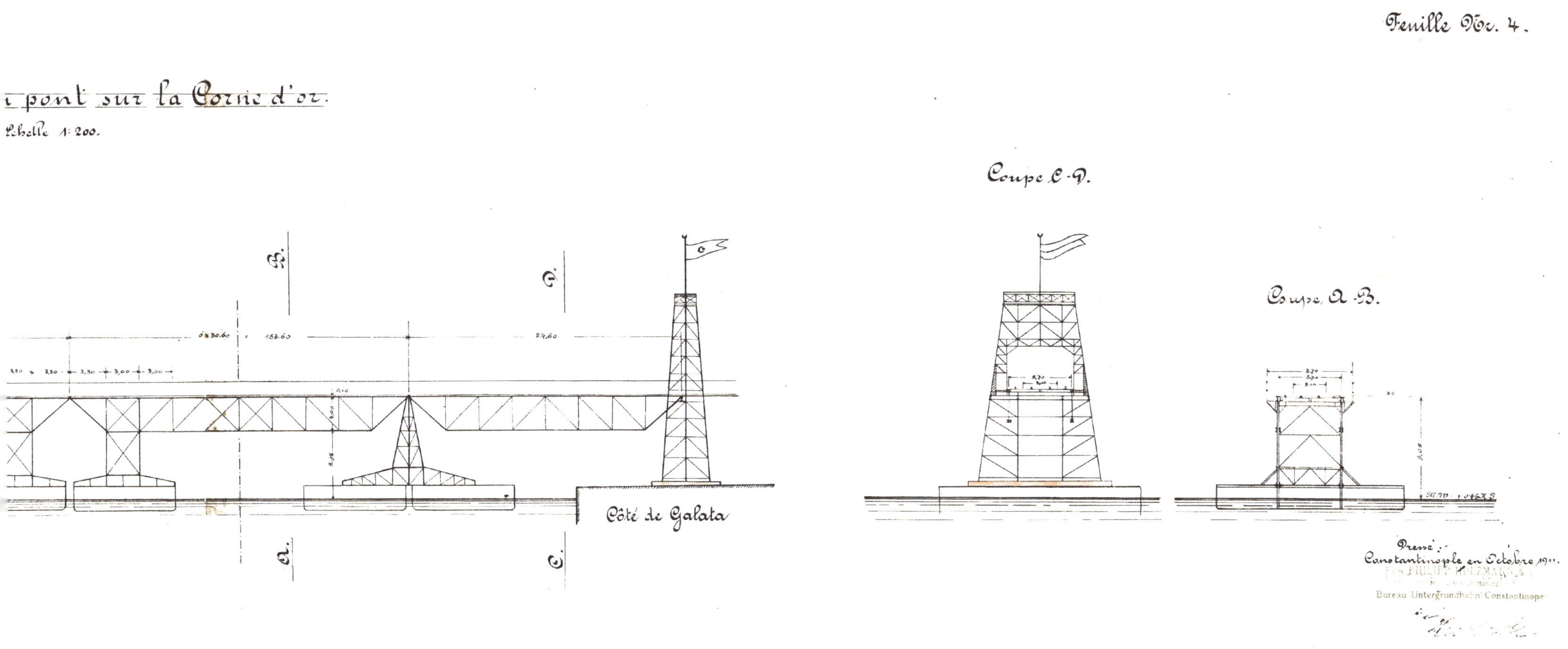

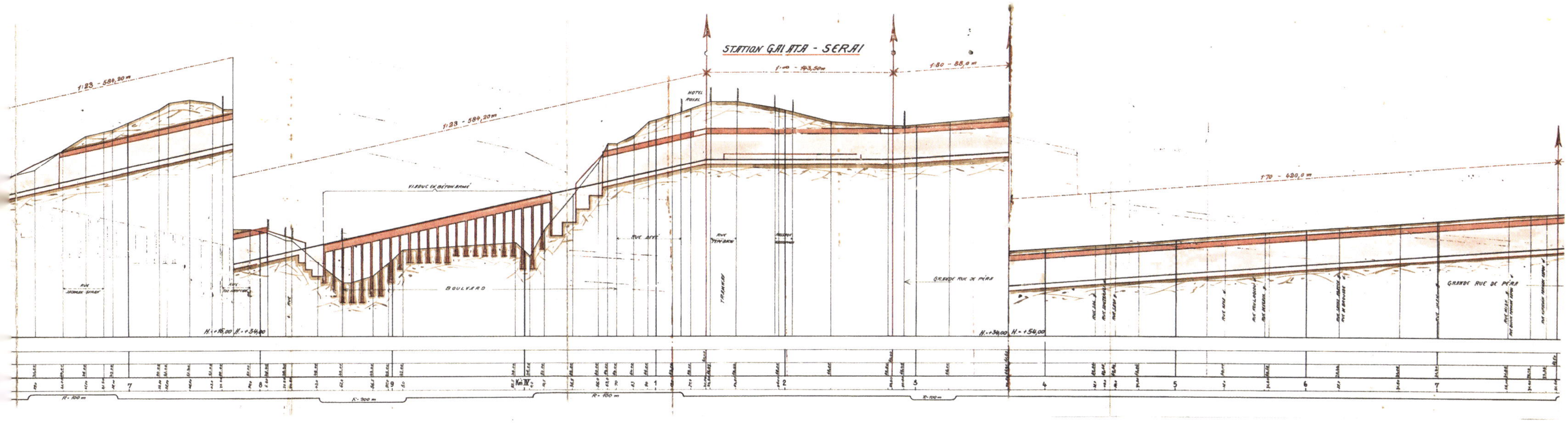

Sections from Philipp Holzmann's Şişli - Beyazıt Metro Project

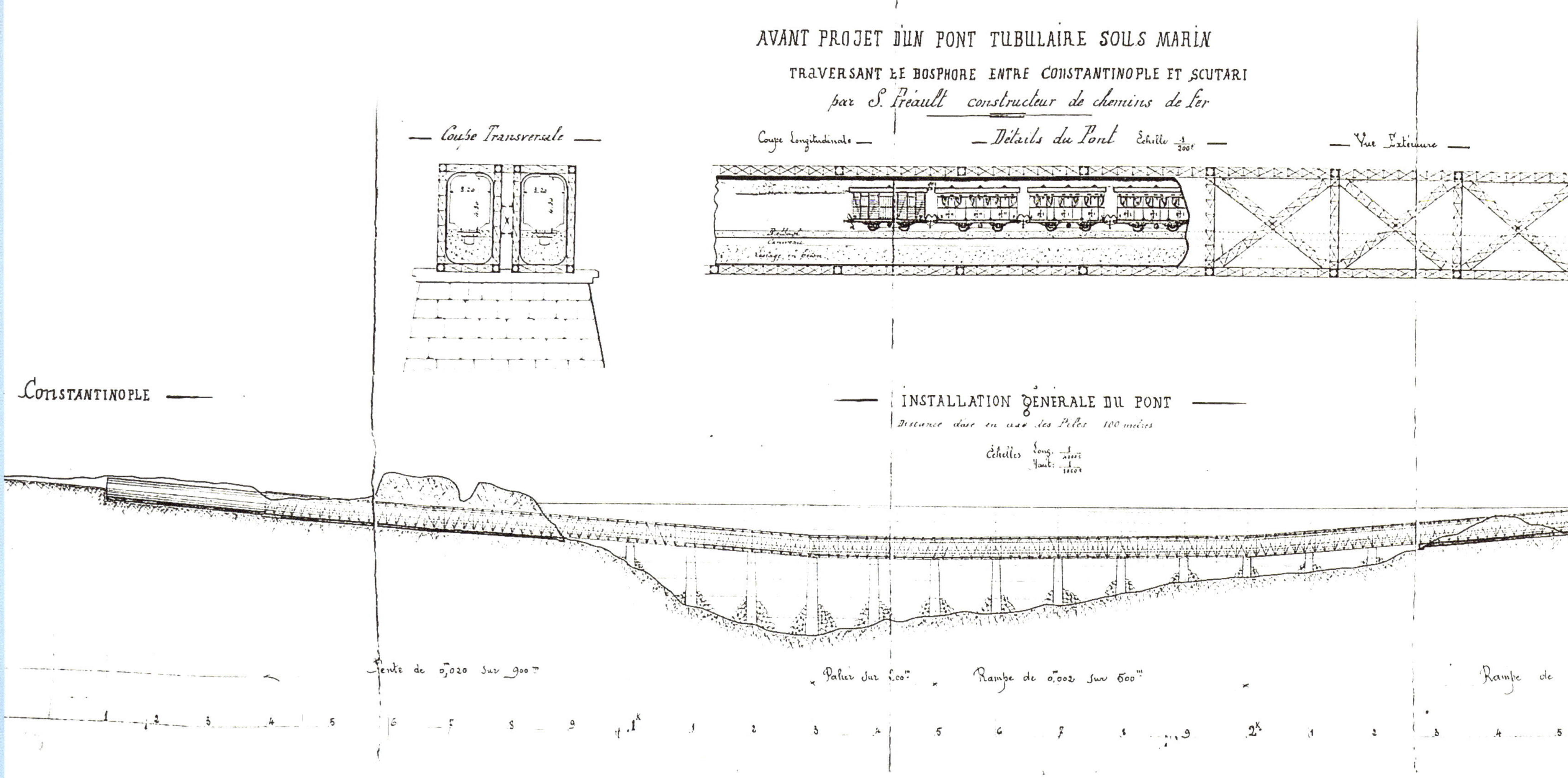

Undersea Tunnel Projects

Simon Préault's Cisr-i Enbubi (Undersea Steel Tunnel) Project

The combined 337 km-long İstanbul-Edirne and Kırklareli-Alpullu parts of the Şark Railways Lines, which fall within the national borders, were completed and put into service in 1888, during Sultan Abdülhamid II's reign. This linked İstanbul to Europe by railroad.

On the Anatolian side of İstanbul, the 91 km-segment of the Haydarpaşa-İzmit was finished in 1873. İstanbul was connected to İzmit in 1873 and then, through the efforts of Sultan Abdülhamid, to Northern and Eastern Anatolia.[76]

The result was that at two ends of the straits, there were two separate "final stations"—one in Sirkeci and the other in Haydarpaşa. The main goal of administrators at the end of the 19th century was connecting these two stations together on a continuous rail line. The obstacle confronting them was the Bosphorus, with its strong currents. The first proposal to link Sirkeci and Haydarpaşa, the two starting points of the Rumeli and Anatolian railroad, came from Simon Préault. The existence of the project was first made known in 1990 in a work entitled *İstanbul'un Eski Haritaları* (Old Maps of İstanbul), written by Cahit Kayra.[77] The sketch, found in the

76 Özyüksel, Murat, *Anadolu ve Bağdat Demiryolları*, İstanbul, 1988, pp. 15–17

77 Kayra, Cahit, *Eski İstanbul'un Eski Haritaları*, İstanbul: İBB, 1990, p. 65

15 - Simon Prerault's Salacak - Sarayburnu Cisr-i Enbubi (Underwater Tunnel) Project - 1891

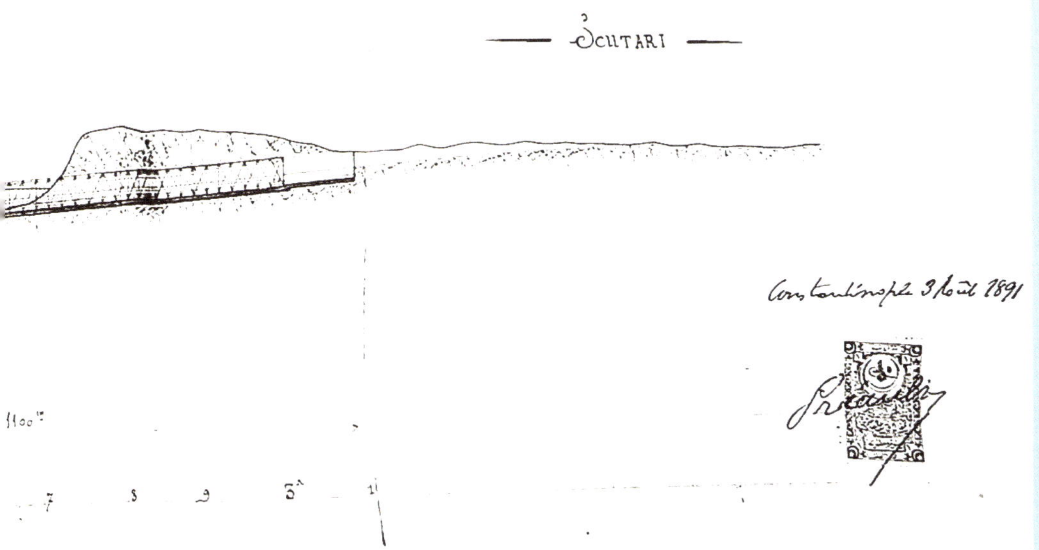

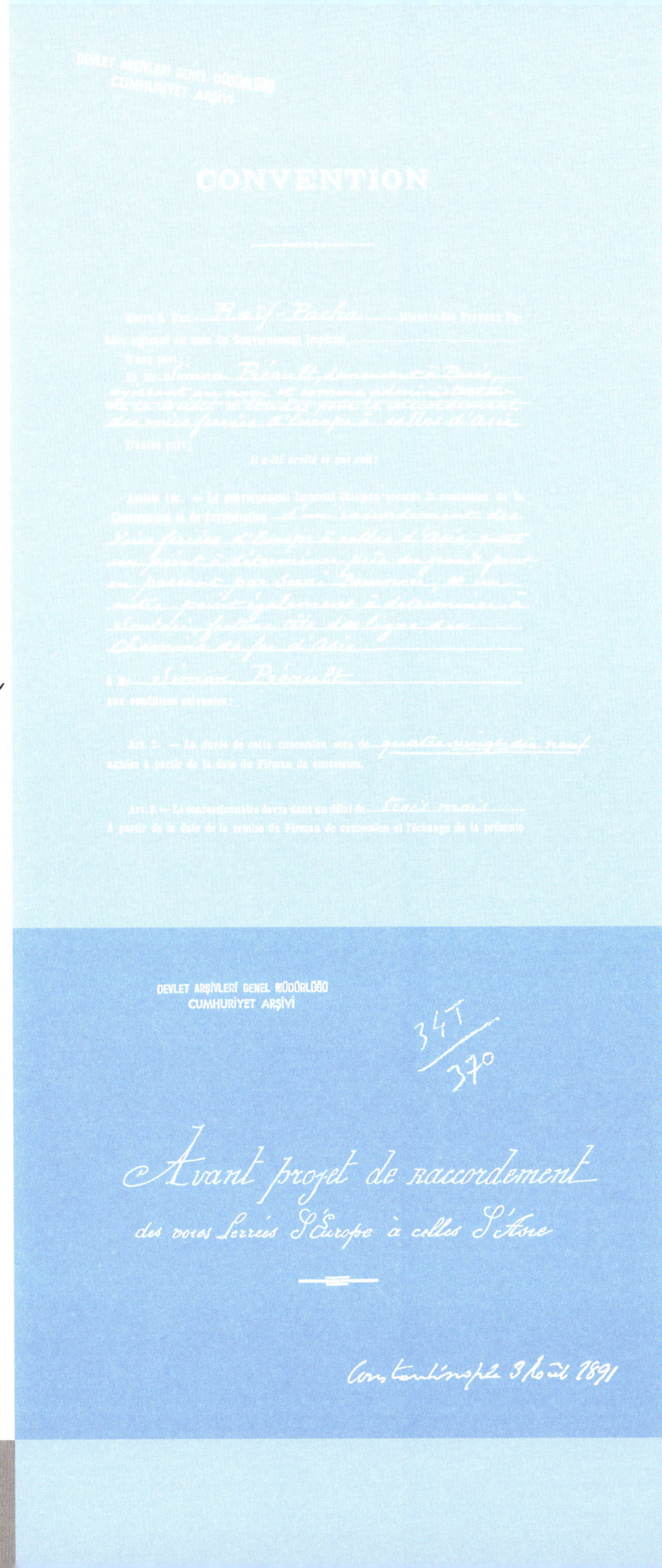
DEVLET ARŞİVLERİ GENEL MÜDÜRLÜĞÜ
CUMHURİYET ARŞİVİ

CONVENTION

Entre S. Exc. Raif-Pacha Ministre des Travaux Publics agissant au nom du Gouvernement Impérial,

D'une part,

Et M. Simon Préault

D'autre part,

Il a été arrêté ce qui suit :

Article 1er. — Le gouvernement Impérial Ottoman accorde la concession de la Construction et de l'exploitation

à M. Simon Préault

aux conditions suivantes :

Art. 2. — La durée de cette concession sera de quatre-vingt-dix-neuf années à partir de la date du Firman de concession.

Art. 3. — Le concessionnaire devra dans un délai de trois mois à partir de la date de la remise du Firman de concession et l'échange de la présente

DEVLET ARŞİVLERİ GENEL MÜDÜRLÜĞÜ
CUMHURİYET ARŞİVİ

34T/370

Avant projet de raccordement

des voies ferrées d'Europe à celles d'Asie

Constantinople 3 Août 1891

Atatürk Library, is undated,[78] but Kayra dates the project as coming from the 1860s, given the drawing of trains traveling in an undersea tunnel.

That the project is actually a work from the Sultan Abdülhamid period and not, at it had been supposed, from the Sultan Abdülaziz period, was confirmed by the discovery of sketches in the Republican Section of the Ankara State Archives. In fact, the black and white sketch was Simon Préault's proposal and it was dated 3 August 1891. The signature it bore was that of Sultan Abdülhamid.[79]

78 İBB Atatürk Kütüphanesi, Harita Kataloğu, 3064

79 Devlet Arşivleri Genel Müdürlüğü Cumhuriyet Arşivi, file 34T370, 230/95.40.2

16 - First presentation of Simon Prerault's Salacak - Sarayburnu Cisr-i Enbubi (Underwater Tunnel) Project - 1891

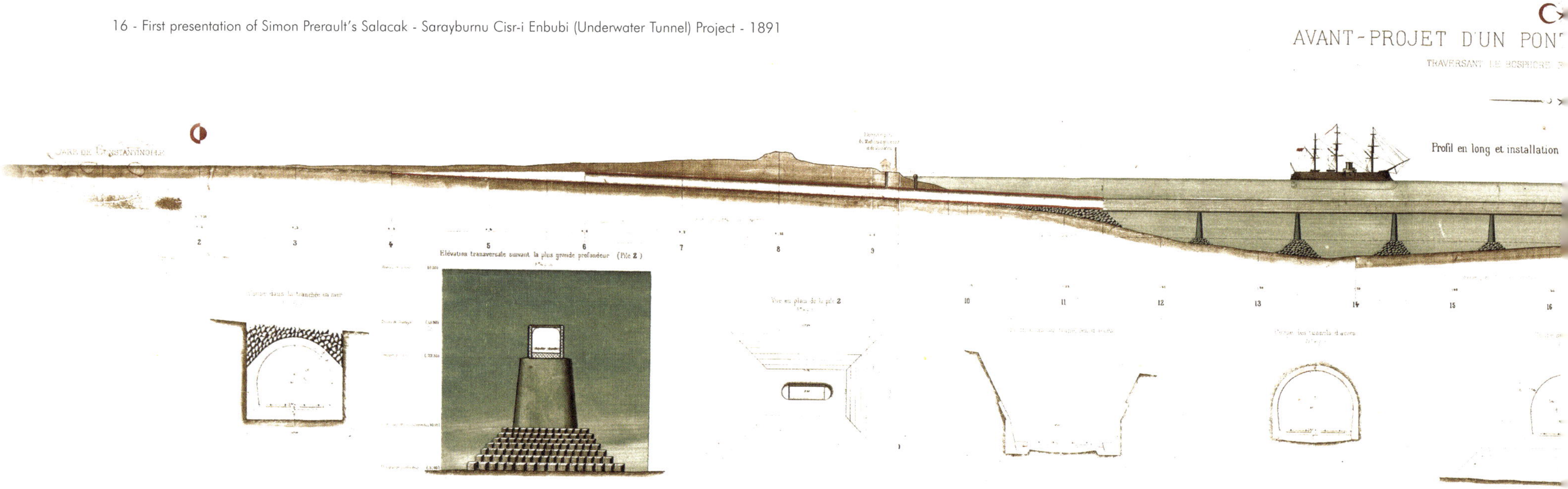

There are technical differences between the undated project, to which color had been added, and the black and white sketch. The computations on each vary, too. Because of the discrepancies between the two, it appears that engineering etudes were done on the project and the original design was improved upon by changing some of the calculations. The most prominent difference is that in the undated one the number of supports is 7, but in the dated sketch, it is 13; moreover, the length of the tunnel in the former was 2,200 m while it was longer—3,100 m—in the latter. In the undated sketch, there is a single tunnel divided into two opposing tracks, but in the dated one, there are two separate tunnels, one on top of the other. The term "Tunnel Passageway" is common to both presentations. In the black and white sketch, each of the tunnels has a height of 4.30 m and a width of 3.20 m. In the colored sketch, the height is 4.40 m. The structure of the iron latticework surrounding the tunnels has been modified on the black and white sketch. In addition, that the drawing of the train inside the tunnel is the same in both sketches shows that there dates are the same.

Nevertheless, in 1891, for an engineer to contemplate such a project and develop it into a proposal, and the administration to accept such a proposal and consider it took audacity. The project was first proposed, without going into technical detail, by Gavand in 1876. That it was brought up once again 15 years later was an indication of the desire to link the two sides of İstanbul. It was interesting that it was not a bridge, but rather a "Cisr-i Enbubi" that was proposed.

Contemporary engineers are of the opinion that, at it appears in the technical drawings, with the steel tunnels supported by 13 buttresses, the project Simon Préault refers to as a "Undersea Steel Tunnel" is not feasible. The Ottoman administrators, who understood the Straits, must have reached the same conclusion since the project was aborted. This was because buttresses resting at the bottom of the sea would never have been able to withstand the strong floor currents of the Straits. On the other hand, today, tunnels are placed upon current-resistant, impermeable structures created at sea level and then covered. But there are still very many similarities between the earlier project and the one being implemented today. In both projects, steel tunnels are being used. In addition, the tunnels are made on ground and then lowered onto the floor of the sea or placed upon buttresses. For each, Salacak-Sarayburnu was thought to be the idea transit route. Having two tunnels and a rail system was another similarity. The total length of Simon Préault's project in his plans was 3.1 km. The tunnel extends across the Straits under water for 1900 m. There is 22 m between the tunnel and the surface of the sea. The lengths of the buttresses that are signed to safeguard the horizontal surface

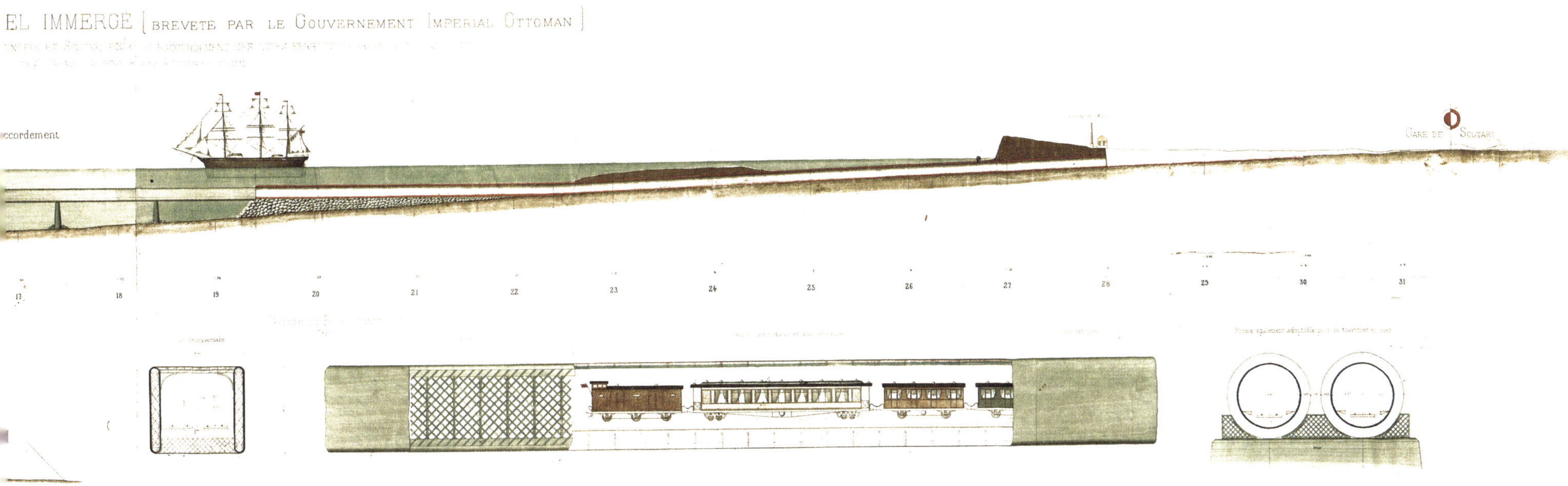

are different. With this type of tunnel the difference between the tunnel and the floor can vary anywhere between 13 and 33 m in places. The connections between the undersea part of the tunnel and its entrance/exits on land have been placed on filled land that reaches 20 meters in spots.

The plans of the project and a proposal letter were sent on 1306 AH to the then Minister of Commerce and Agriculture, Raif Pasha.[80] However, it is interesting that the proposal, which had been archived in the same file as other bridge and undersea tunnel projects submitted at later dates, had two covers—one dated 1307 AH and the other 1891. The text of the proposal was in French and there was no Ottoman translation in the file. The proposal was 33 pages long and divided into sections: Franchising Agreement, contract, construction project, maintenance and operation, and company. The proposal also had a number of blanks that needed to be filled in and in places there are sentences that have been crossed out. Therefore, it can be thought that the proposal was a preliminary text. Under the heading "Construction Project," apart from technical information, for example, that there would be a 20 mm-slope for one meter at the tunnel's entrance and exit, there are no interesting articles. From Article 21: "Historical artifacts found during construction will be reported and handed over the Ottoman State." It should be asked if the reaching of İstanbul's oldest harbor during the excavation of the current undersea tunnel's Üsküdar Station was anticipated by Article 21. Then there is Article 19, which says "Except for technicians, all other foreign laborers shall wear a fez.

11 years after Simon Préault, another undersea tunnel project was called to the attention of the Sublime Port. The project of the American engineer Frederik E. Strom and colleagues was translated into Turkish as "*Cisr-i Enbubi fil bahr*" and submitted to the Sultan.[81] The vehicle through which the presentation was made was one of the most powerful admirals of the day, Ahmed Besim Pasha.[82] İsmail İşmeni, who first published the projects in 1972, says that "even the franchise for the water tunnel between Salacak and Sarayburnu was obtained."[83] The Sultan most likely wanted to honor the bold engineers. The proposal of Strom and his colleagues called for establishing uninterrupted railroad connection between Europe and Asia. However, it contained technical omissions and lack of detail in the drawings. Even Simon Préault's proposal submitted 11 years earlier seemed better.

A three-car train, two for passengers and one for belongings, would operate inside a tunnel restrained by 16 buttresses on the bottom of the sea. The location of the stations was not fully established in the proposal. The project had the idea of a monumental station on the Asian side of the city across from the Sirkeci Train Station, which had been built in 1890. The government understood that this project had too many technical deficiencies to make it feasible.[84]

80 Devlet Arşivleri Genel Müdürlüğü Cumhuriyet Arşivi, File 34T370, 230/95.40.2, No 3

81 İstanbul University Rare Works Library, No. 93235

82 Kayra, *op cit.*, p. 65

83 İşmen, İsmail, *İnsanlar ve Köprüler*, İstanbul: İş Bankası Kültür Yayınları, 1972, p. 466

84 İlter, İsmet, *Boğaz Geçişlerinin Tarihçesi*, Karayolları Genel Müdürlüğü, 1973, p. 24

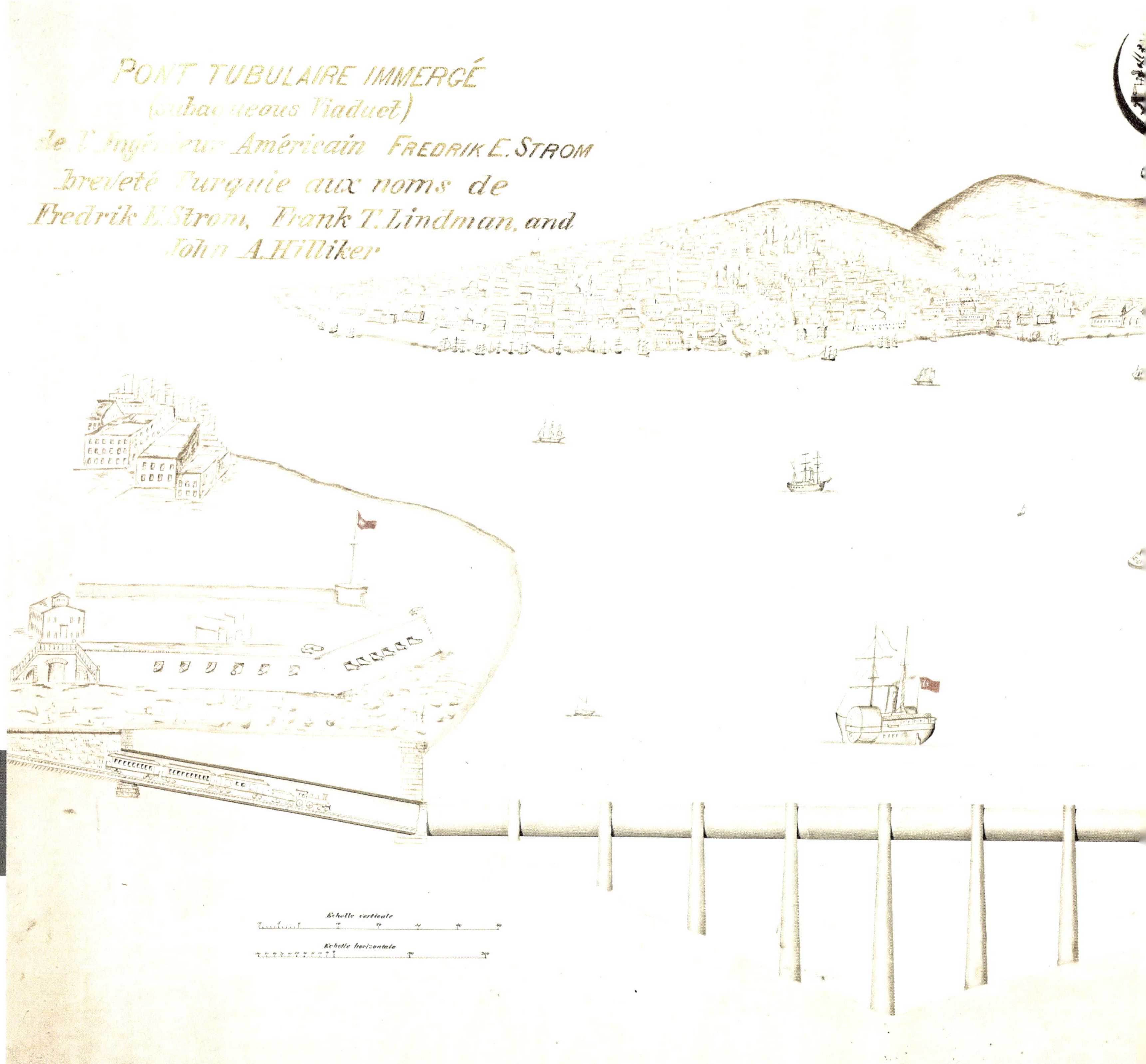
PONT TUBULAIRE IMMERGÉ
(Subaqueous Viaduct)
de l'Ingénieur Américain FREDRIK E. STROM
breveté Turquie aux noms de
Fredrik E. Strom, Frank T. Lindman, and
John A. Hilliker
Echelle verticale
Echelle horizontale

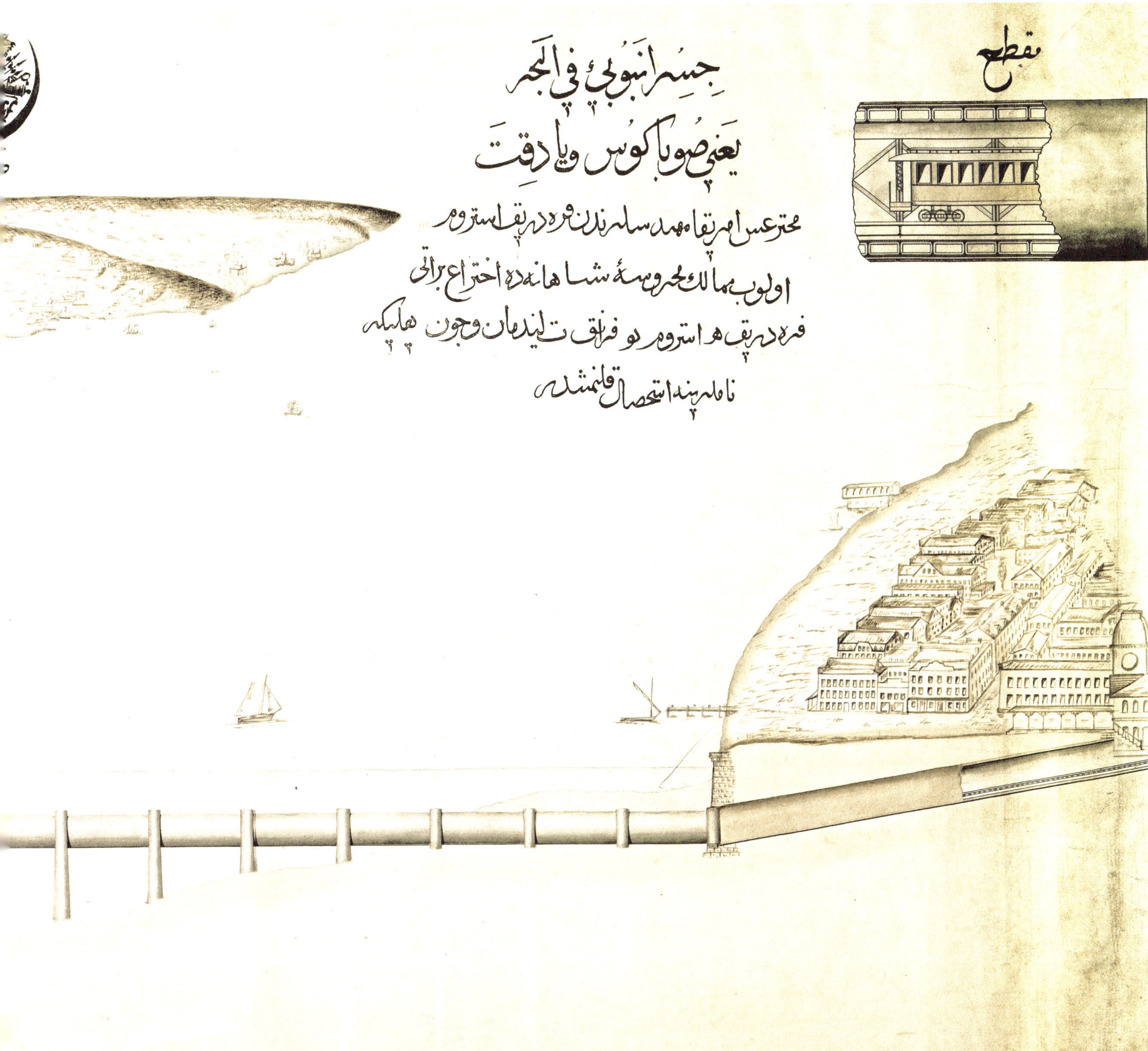

17 - Frederik E. Strom and colleagues' Salacak - Sarayburnu Tünel-i Bahri (Underwater Tunnel) Project - 1902

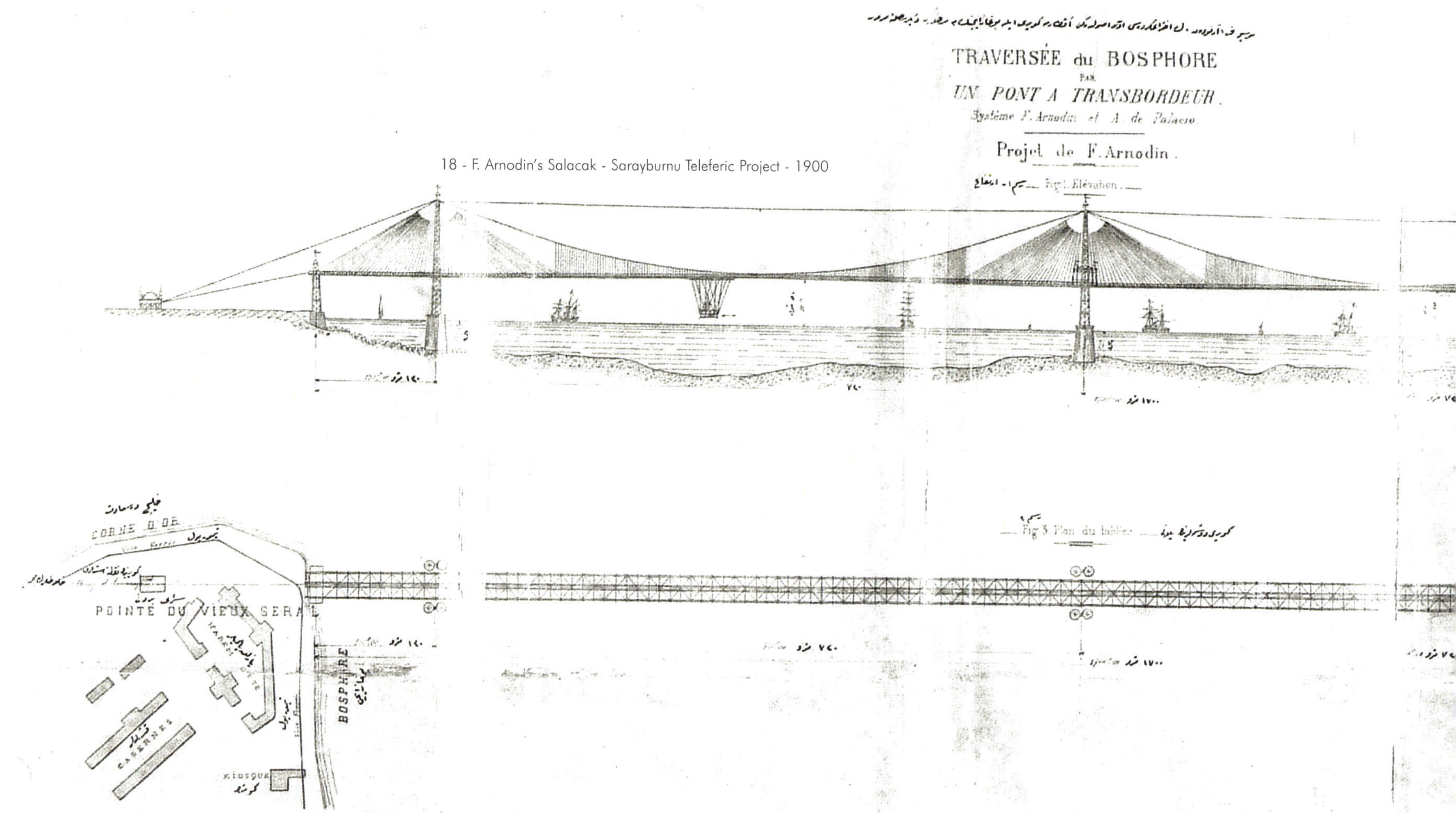

18 - F. Arnodin's Salacak - Sarayburnu Teleferic Project - 1900

F. Arnodin's Cisr-i Hamidi (Hamidiye Bridge) and Freeway Project

It is rumored that the first pedestrian passageway on the İstanbul Straits was built before the Common Era. Persian King Darius, seeking to expand the borders of his sovereignty westward, reached the Straits after having advanced upon the Scythians. Prior to the military expedition, Darius had commanded his army to construct a bridge to go over the Thracian Straits. When he got to the Straits, he found the bridge made by Mandrocles, from Samos, completed. In remembrance of the bridge, Darius had two columns raised and had the names of the nations that had participated in the expedition engraved on them, in Assyrian on one and Greek on the other. Later, these columns would be used by the Byzantines as the altar in the temple of Diana of Orthosia.[85] Again, according to what Herodotus, the second column, which was inscribed in Assyrian, was used by Byzantium near the Temple of Bacchus.

There is no technical information about how Mandrocles' bridge was built. What we do know is that it was a pontoon-type made between Rumelihisarı and Anadoluhisari, the narrowest places of the İstanbul Straits, in order to line ships and ferries next to one another.[86] The armies of Alexander passed the Straits in 334 BCE, those of Abd al-Malik in 717, and of Orhan Bey in 1352, but the methods they used were not the same. These crossings could be made by ship or small boats especially designed for this crossing.

Apart from the bridge Da Vinci designed for the Golden Horn in 1503, and his proposal that it could be extended into Anatolia, the first serious initiative to construct a bridge over the Straits came from Ferdinand Arnodin. The map of the feeder line route and a plan of the bridges

85 İlter, *op cit.*, p. 20

86 Rowlinson, George, *Heredot Tarihi*, İstanbul: Kanaat Kitapevi, 1941, v. 4, p. 87

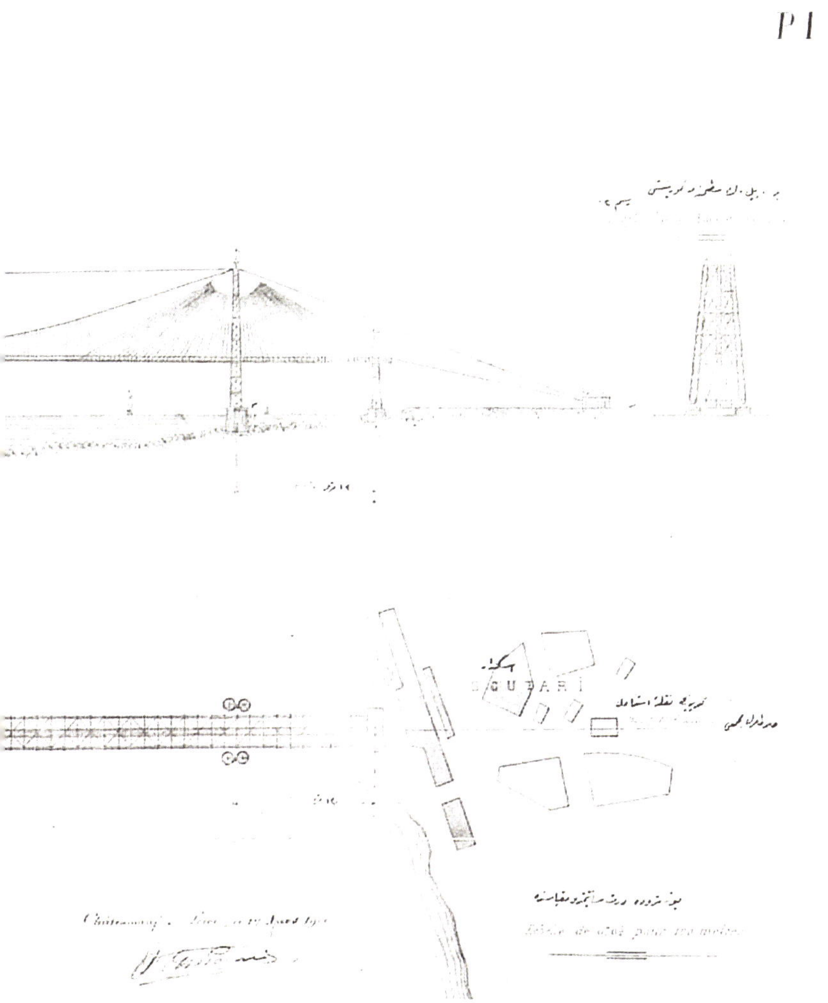

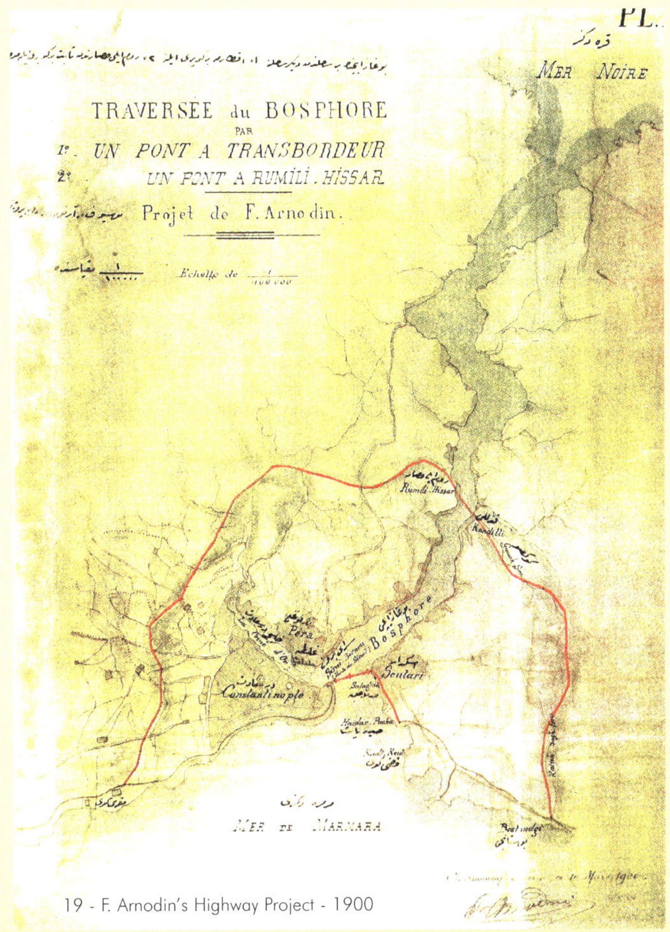

19 - F. Arnodin's Highway Project - 1900

were submitted to the Sultan, along with an explanation written on behalf of the company Compagnie Internationale du Chemin de Fer du Bosphore, in March 1900[87] The main objective of the project was providing a railroad link between Europe and Asia. However, the organization of pedestrian and vehicle traffic was also part of the project. In this way, this project became more than a new railroad project and was transformed in an urban, perhaps even regional plan.

The railroad that was to pass over the Hamidiye Bridge, which was to be constructed between Rumeli and Kandilli, would incorporate the Bakırköy and Bostancı Stations. It was thought that İstanbul would develop in this direction in subsequent years, which is why the line was kept extensive.

The impressive size and unusual architecture of Arnodin's bridges would add striking new elements to the city's silhouette. In particular, the telpher running between Sarayburnu and Üsküdar would resemble a monumental gate at the entrance of İstanbul. This structure, which was to be in the form of a suspended telpher, would sit on two buttresses, each 130 meters from the shore. Almost void of embellishment, above all, it was thought of as an engineering marvel. Nevertheless, it was not completely without new motifs: tiny domes were affixed atop the support buttresses, which made it resemble mosques with minarets. Moreover, two domed mosques, each with two minarets, at both points where gondolas coming and going reach the shore, were being considered.

87 BOA, YA.HUS, 411/174

What distinguished Arnodin's project from that of others was the uniqueness of the bridges. Neither Préault nor Strom had bridges in their projects that resembled the ones in Arnodin's. Indeed, it might well be the first idea of an undersea tunnel project. The telpher-style bridge (Transporter Bridge) Arnodin proposed be built between Salacak and Sarayburnu had been introduced in Portugalete in 1893, Raven in 1899, Bizerte in 1898 and Rochefort in 1900. Another five of this type of bridges were to be built after 1900. But one wonders how construction of a bridge of this kind between Salacak and Sarayburnu could be considered without taking into account the actual requirements of the bridge. While what was needed between the Sirkeci and Haydarpaşa Stations was a continuous railroad, what Arnodin had in mind was not a bridge linking the two stations, but rather a one-way transportation system operating with gondolas.

The second bridge, which was much more architecturally ambitious, would perform this function and connect the two sides by a continuous railway. This was a suspension bridge supported by three buttresses. In contrast to the telpher, made with iron and steel, with its relatively dynamic appearance, at the entrance to the Straits, the Hamidiye Bridge, bearing the name of the Sultan, evoked more romantic associations. The buttresses of this bridge were transformed at their base into mosques. Each mosque had a central dome and four minarets. According to the description submitted with the project, the architecture of the bridge consisted of both Seljuk and other Islamic styles. The domes resembled the palaces of Islamic caliphs and exalted the power of the Ottoman Sultan, who was the Muslim caliph. The bridge was embellished in Arabesque style with colored bridges, tiles and shiny metals. The Hamidiye Bridge, displayed all of the beauty of 16th and 17th century Northwest African architecture, was even more enchanting at night. Moreover, the bridge would have yet another interesting function. In the text of the submittal, it was stressed that the design would have a defensive purpose. Each one of the high domes is to be built upon a granite base, surrounded by large cannons, revolving towers, and an undersea platform impervious to attack.

What was submitted to the Sultan was not only plans for the bridge and the map of the peripheral roads leading to it. There was also a text

that had been translated by the Ministry of Interior Translation Office. The text illustrates how a project was proposed in the 1900s and also provides a detailed description of the bridges. Therefore, it is worth reading.

"The idea of building a bridge over the Bosphorus was discussed before the Russo-Turkish War (1877). This time, with the Anatolia-Baghdad Railroad project, the franchise of which had been awarded the Germans and was going to start soon, this idea once gained

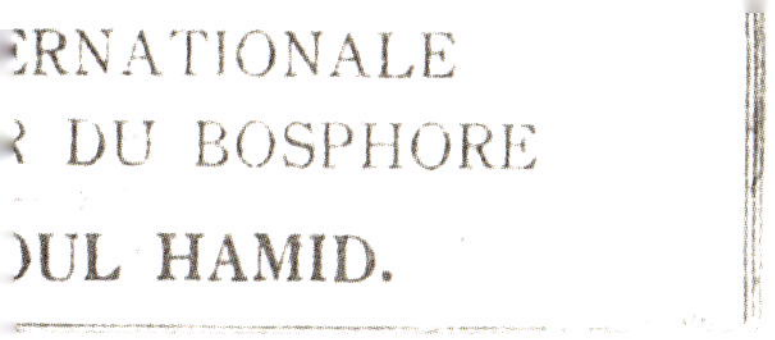
ERNATIONALE
R DU BOSPHORE
DUL HAMID.

20 - F. Arnodin's Anadolu Hisarı - Rumeli Hisarı Hamidiye Bosphorous Bridge Project - 1900

importance. The idea of a Bosphorus Railway Company came from linking railroads in the south and southwest of Europe to the Baghdad Railroad with a large bridge that would cross the Bosphorus at its narrowest point: between Rumelihisarı and Anadoluhisarı. Even the military bridge over which Darius passed with 800,000 Persians in the 6th century BCE, built by the Corinth Architect Mandrokal and considered one of the most beautiful famous works in the world, was built at this point on the Straits.

A picture shows what this big bridge would look like after having been completed. The distance between the two coasts at the point mentioned is approximately 600 meters. This great work, with its embellished minarets and domes elaborated done in Northwest African architectural style, and steel cables situated among the solid stone buttresses, gave it an elegant ethereal appearance of iron latticework.

Each of these high domes are built upon a granite foundation, surrounded by large cannons, and protected by revolving towers. The whole structure sits upon an undersea base impervious to attack. The Hamidiye Bridge, with its three suspension bridges, was to provide quite a statuesque vista. In addition, the elegant appearance created by the high domes and bell towers surrounded by 15 lanterns will bring it that much closer to perfection. When the thousands of electric lights radiating around the bridge are turned on at night, the fine splendor of the arabesque ornamentation will be projected. The bridge will be blanketed with beauty from one end to the other and become a focus of unquenchable desire. The craftsmanship of this gorgeous work, which will be made of iron and stone, was inspired by the elegant architectural style of old Seljuk buildings. And just take a look at these marble domes, which are reminiscent of the places of the caliphs of Islam. They will exalt before the eyes of everyone the whole of the power of the religious and political power of the Sultan, who is caliph of the Muslims. It would manifest the glory and grandeur of the Ottomans. Look with amazement and admiration at these gates in the middle of the bridge made for the trains to pass through. With Its enormous domes, surrounded on all sides by minarets and ornamented with colorful bricks, tiles and gilded bronze, it embodies all the characteristics of 16 and 17th-century Arabesque-style Northwest African architecture. The shadow of the suspension bridge, which resembles lace strung between these splendid towers, falls beautifully upon the radiant eastern sky. At first glance, we Venus see, the ship that belongs to the Lloyd Company, pass under the iron niche of the bridge. Further ahead, can be seen the Anatolian side of the Bosphorus and the while houses in Anadoluhisarı. This Hamidiye Bridge and railroad network, over which trains, cars and pedestrians pass, extends in two directions, to İstanbul and Üsküdar.[88]

88 BOA, YA.HUS, 411/174

The Golden Horn Bridge Projects

Historical sources mention a wooden bridge having been built at the narrowest point of the Golden Horn in the 5th century. Emperor Justinian replaced this in 528 with a stone bridge, but unfortunately it did not remain standing until the Ottoman era. In the meantime, transit from Karaköy to Eminönü was achieved by primitive means.[89]

(Fatih) Sultan Mehmed II had a bridge built during the siege of İstanbul by lining up ships side by side to launch an attack on the city wall from the sea. He later had them removed after the conquest. Despite all the difficulties involved in it, the idea of building a bridge to the Golden Horn remained in minds and many projects were developed for this purpose.

The first proposal was submitted to the Sultan in 1503. Leonardo da Vinci developed a bridge project that would connect Pera with İstanbul over the Golden Horn. His plan consisted of a single-arched bridge 240 meters in length. The proposal letter Da Vinci sent to Sultan Bayezid was written on 3 July and a facsimile of it was prepared and submitted to the Sultan four months later, in October.[90] As can be understood from the heading "This is a copy of the letter sent by the Linardo from Genoa," it did not simply contain a proposal for a Golden Horn Bridge. Among the other projects proposed was a mill powered by only by wind and not water, a new water wheel to draw water from ships, a bridge in Galata and, if requested, a bridge going over the Straits to Anatolia. In the letter that proposed the bridge, Da Vinci had this to say:

> I, your servant, have thought about the mill and, through God's help, I have a way of building a mill that operate on wind alone, without the need for water. Fortunately, God has ordained me to find a way of drawing water from ships without using cords or ropes, and only using a hydraulic machine that runs on its own.
>
> I, your servant, have heard that you want to build a bridge from İstanbul to Galata, but haven't been able to find anyone who can do it. I, your servant, know how it can be built. I'll build a bridge as high as a building. Because it is very high, no one will object to crossing it. I'll make such a bridge that even a ship with its sails unfurled will be able to pass under it. I'll build a suspension bridge whose span will reach the Anatolian coast, for anyone requesting one. May God have you believe these words. Know that this servant of yours will always be at your service...
>
> This letter was written on the 3rd of the month of Julius...[91]

On the other hand, among notes of Leonardo preserved in the library of the Institute of France in Paris, there is a simple diagram of the bridge designed for the Golden Horn. The explanation under this plan was written backwards and could be read by looking at it through a mirror: "bridge from Beyoğlu to İstanbul; a width of 40 archines (30 meters), a height of 70 archines (52 meters) above water and a length of 600 archines (450 meters), of which 400 archines (300 meters) will be self-supporting on land." Prof. F. Stuss, who has studied the preliminary project, expresses his concern about the design, saying: "From Leonardo's perspective, was it unfortunate that he was unable to get the bridge built, or was it just the opposite? Was he saved from having his great fame overshadowed by failure, or was the world deprived of the construction of an inimitable bridge? These are questions to which I am unable to provide an answer."[92]

89 Çelik, *op cit.*, p. 72

90 Topkapı Sarayı Archive, Document No 6184

91 Eyice, Semavi, *Belgelerle Türk Tarihi Dergisi*, issue 19. p. 24

92 İlter, *op cit.*, p. 52

21 - Da Vinci's Galata Bridge Project - 1503

93 "Da Vinci Köprüsü" (documentary), Cengiz Özdemir and Adem Özkul, TRT, 17 January 2009

94 İlter, *op cit.*, p. 53

95 İsmail, Namık, *Mikelanj "Michel-ange,"* İstanbul: Milli Eğitim Bakanlığı, 1931, pp. 11–12

Meanwhile, 30 km south of Oslo, on the inscription of a bridge built in the town of Aas by the architect Veb'yurn Sand, using Leonardo's plan as the basis for its construction, it says:

Leonardo da Vinci had a vision in 1502: to build the largest and most beautiful bridge the world had ever seen. The bridge was to span in an arc-like fashion for a distance of 250 m from Constantinople to Pera. But Sultan Bayezid II, who had ordered the project, panicked over the dimensions of the project and the bridge project was aborted, never to be implemented. The letter Leonardo wrote to the Sultan was uncovered in the 1950s, and the thought that the bridge could be built got everyone excited. The basic lines of the draft Leonardo drew 500 years earlier proved that his art was timeless. The bridge represented the coalescence of Renaissance science, art and humanist thought, in one sense, the merging of the ethereal and temporal.

Architect Veb'yurn does a disservice to Sultan Bayezid II by saying that he "panicked over the dimensions of the bridge." It is quite natural for the claim of building a single-arched bridge of 240 m in length made in a one-page letter unaccompanied by any drawing or plan from an artist who had yet to prove his reliability, under the conditions existing at that time, to have been rejected by the Ottoman palace. What's more, the same proposal included a single-arched bridge to span the Straits as well.

It would not be far from the truth to say that this project would stretch even the limits of today's advanced construction technology. Nevertheless, Prime Minister Recep Tayyip Erdoğan announced that despite all the difficulties, they wanted this project to be carried out. In a statement he made in a documentary filmed in 2009 entitled "Da Vinci's Bridge," he announced that what had been an Ottoman dream was becoming a reality, saying:

It's thrilling that such a bridge is going to be built between Sütlüce and Eyüp Sultan, and that the idea has become a reality, whether as part of the work being carried out by the Greater Municipality of İstanbul or as part of the 2010 European Culture Capital project. This project will unfold in a very meaningful place—the Golden Horn. Of course, it's a question of a significant contribution to universal values. The construction here of a work of someone the likes of Da Vinci, who has left a lasting historical legacy, will serve as a building block in the integration of Turkey into Europe. I congratulate everyone who has a hand in making this a reality. Above all, I wish everyone involved in both the architectural legwork and the implementation the greatest of success. I believe that this bridge will make a very distinct contribution, enabling anyone passing over it to get to known Da Vinci more intimately.[93]

Historical sources reveal that there was another initiative put forth to construct a bridge on the Golden Horn three years after Da Vinci's made his. This time, it was by Michelangelo.

The greater master, who had had a falling out with the Pope toward the end of 1505, flees Rome and goes to Florence. Frightened upon hearing that the Pope would excommunicate him, the Seigneur of Florence tries to convince the artist to return to Rome. According to the painter Ascanio Condivi, who wrote about the life of Michelangelo, the artist considers going to Turkey[94] Michelangelo is told by a Franciscan monk that the Sultan wants to have a bridge constructed from Pera to İstanbul.

Another biographer of Michelangelo, the Russian Dimitri Merockovski, provides detailed information about the subject:

The idea of building a bridge in İstanbul was attractive to the great artist. Wishing to escape from the Pope, he had gone to the home of a Franciscan monk. As usual, the monk accepted him joyously and offered him rose jelly and deserts. Showing him letters he had received from İstanbul, he said that as soon as he made up his mind, he could go to İstanbul on Turkish ships already in the harbor.

Michelangelo asked,

"What would the Sultan want from me? To be a Muslim?" The old monk responded "Rest assured, my son," and continued, "The Sultan will not ask you to do anything against your conscience. I have lived in both Rome and İstanbul. I can't tell which city has more sinners. I saw such Muslims who were more upright and performed more good deeds than many Christians, who, holding their prayer beads would have a steady stream of insincere prayer falling from their lips.

Michelangelo persisted, saying

Will I have freedom? Will the Sultan allow me to do whatever I want in the area of art? "Listen, my son; during the time of the Greeks, a sculptor wanted to make a sculpture for the city of Alexandria so great that the entire city—its houses, squares and streets—would fit into its palm. If you have the courage to do something on a grander scale than this, only in İstanbul will you be able to do so. The sultan—a thousand times more powerful and richer than the Pope—desires the construction of something the magnificence of which no mortal has yet to imagine.

Hearing that Michelangelo was about to leave for İstanbul, the Pope immediately penned a letter, saying that if he went to the Sultan, he would be excommunicated. He wrote that he would be well accepted if he returned to Rome. Hence, Michelangelo's plan to go to İstanbul fell through."[95]

Nevertheless, this was not the end of the idea of bringing Michelangelo to İstanbul. Years later, the same plan came up once again. In April 1519, Tommoassodi Tolfo, one of the foreign merchants in Edirne, wrote a letter in which he invited the artist to İstanbul once more. The letter sent to Michelangelo on 1 April 1519 by Tommoassodi Tolfo, the representative of the Florence Gondi Bank in Edirne, is extremely important because it revealed another reason underlying the failure of the initial endeavor.

Writing in his letter "When we met a number of times in the home of Gianozo Salvati, you indicated your desire to come here and see the country; but I convinced you otherwise, telling you that the Sultan at that time did not like art of any kind," Tolfo admitted how he had influenced Michelangelo, who had confirmed what the Franciscan monk had told him 16 years before.

22 - Painting depicting the meeting between Michelangelo and the Turkish Ambassador

96 Eyice, *op cit.*, p. 28

97 İlter, *op cit.*, p. 74

In latter lines, Tolfo is inconsistent when he says, "The position of the current Sultan (Yavuz Selim) is completely different. If you still have the same idea, I recommend that you leave immediately to come here."

Tolfo tells Michelangelo that the Sultan is cognizant of the proposal: "Leave from Raguza. I will assume the responsibility for getting an edict from our Sultan before you reach Raguza. Trust me, either you or someone else come; there will be no dearth of God's blessings"

It is clear that Tolfo's attitude, which was full of contradictions, put Michelangelo into a predicament. Hence, the second initiative remains inconclusive. However, a painting depicting the invitation of Michelangelo to İstanbul can be seen in the home of an artist in Florence that has been turned into a museum, Giovanni Biliverti, who did the painting, is actually a Dutch painter who lived for a long time in Florence.

In Biliverti's painting, Michelangelo is depicted before an Ottoman mission. Apart from a bearded Turkish ambassador, there is a black child and two other Turks who are taking out the edict from the repository carried by the child. According to Semavi Eyice, "Given that Michelangelo had never been invited by a Turkish mission of this size, Biliverti fabricated this painting as an expression of his admiration for the artist."[96]

The search for ways of connecting the two sides of the Golden Horn would undoubtedly continue in later years. The first bridge built on the Golden Horn was in the 19th century—in 1836. The wooden bridge, which was high enough to allow the passage of small boats underneath, connected Azapkapı to Unkapanı. It was roughly 600 m long and 10 m wide.

Galata began to develop rapidly after 1838; this development also increased the need for a facilitation of transportation between the two shores. The first Galata Bridge was built by the mother of Sultan Mahmud II, Bezmialem Valide Sultan, in 1845. The bridge was replaced with a bigger and sounder wooden structure in 1863; it served İstanbul until 1869, when an English company made a new bridge proposal.

The English company proposed a bridge having mostly iron components. Furthermore, large docks at both ends of the bridge were planned. The government approved this project, but the first plans were modified. In order to complete the project, the English company submitted a report to Sultan Abdülaziz in 1871, proposing that the new structure be moved and replace the old bridge between Unkapanı and Azapkapı. The government adopted this proposal and in 1872, the entire structure was moved and replaced the old Unkapanı Bridge. The construction of the Galata Bridge was begun for the second time in 1875 and was completed in 1878. Sultan Abdülhamid, who ascended to the throne in 1876, took an interest in the project and had it developed. In fact, the very first project the Sultan had carried out was this bridge. The new iron structure was 480 m long and 14 m wide. In addition, there were 2.15 m-pedestrian walkways and a 9.7 m-road for cars on the bridge. The bridge, supported by 24 pontoons, was designed to open from the middle to allow the passage of maritime traffic. There were shops, restaurants and coffee shops toward both ends of the bridge. There was even a sea bath on the bridge looking out onto the Golden Horn. Until a new one was built in 1912, this bridge connected İstanbul to Galata. After this date, it was moved to Unkapanı, where it replaced the old one there.[97]

The Sultan's active interest in the construction of the first bridge on the Golden Horn was encouraging to entrepreneurs. It paved the way for many Golden Horn bridge ideas, some of which were approved and for which contracts were drawn up.

Projects for which details are known

The first of these was the 1894 project entitled Plans for the Hamidiye Bridge, which is to be built from stone and constructed between İstanbul and Galata, the blueprints for which Monsieur Antoine Corenty prepared and submitted to receive a franchise for the purpose of building it.[98] The proposal belonged to Antoine Corenty, who was born in 1839 in İstanbul to a family of Italian origins. Corenty designed a bridge to span the Golden Horn; it was to be an eight-arch bridge supported by eight timber columns. They would be positioned, four on the right and four on the left of a suspension mechanism that can be opened and closed to allow the passage of ships underneath. He added some extremely interesting elements to the design.

Corenty divided the surface of the bridge into three, reserving the middle to a rails and the outside of the bridge overlooking the Golden Horn for a row of three-storey buildings. At both ends of the bridge, he placed triumphal arches.

The second proposal came from the Nuremberg Machine Company and was dated 1895. The project contained three different alternatives.[99] However, the plans for the bridge to be built upon buoys were not included. Each of the bridges in the plan could be opened and closed enabling the passage of ships. These bridges were at the same time the first suspension bridges considered for İstanbul. The plans for the first two bridges, both of which had iron components, were quite detailed, as well as sophisticated for the age. The bridge in the first plan was 430 meters in width, with buttresses bearing cables being positioned at 70 meter intervals. Protrusions resembling domed lighthouses were added to the Karaköy and Eminönü ends of the bridge. The structure was topped off with a crescent and star. In the second plan, the buttresses were to be on land and 100 meters in length. The crescent and start was moved to the apex of each of them.

98 İstanbul University Rare Works Library, No. 824

99 BOA, Y.PRK.TNF, file 4, case 21, 23. Şa. 1312

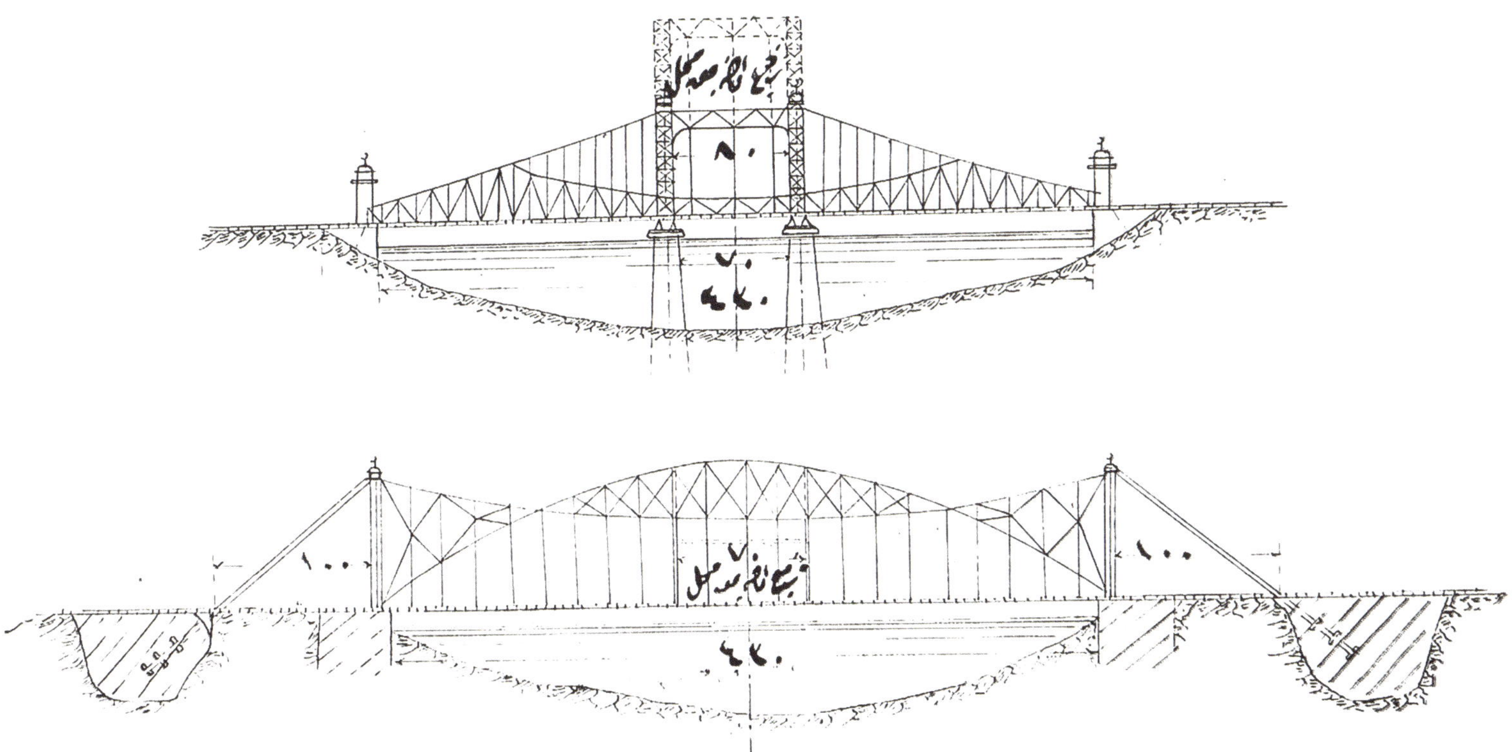

23 - The Gold Horn Project of Nuremberg Machine Company - 1895

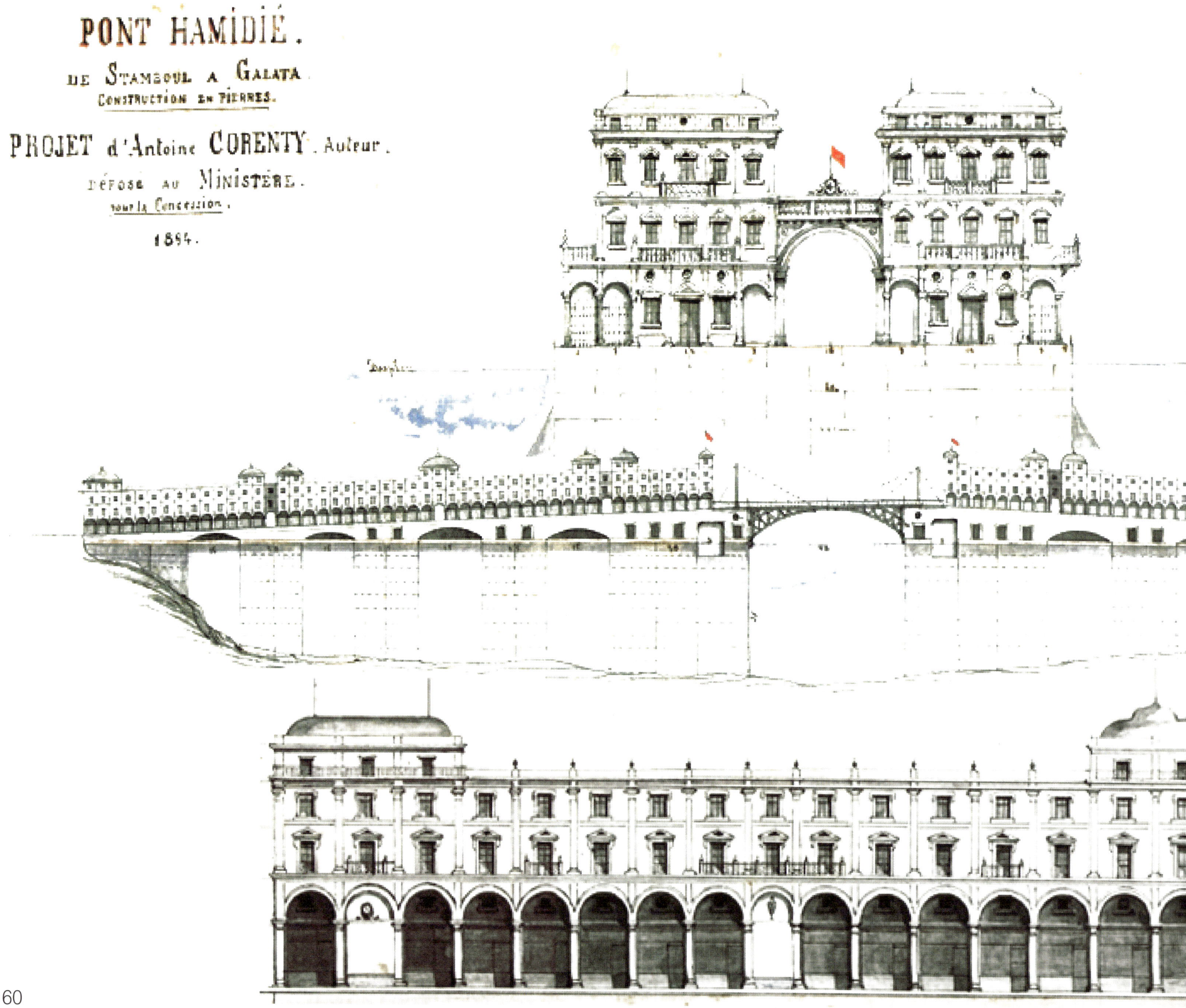

24 - The Golden Horn Hamidiye Bridge Project of Antoine Corenty - 1894

The third proposal was sent from Paris. The project, developed by Monsieur Pierre and dated 1902, included a detail iron frame and a mechanism operating on a series of electrical components. The bridge was to have three parts, with each one punctuated by a pair of minarets. The minarets are connected to one another by balconies. The components of the iron arch of the structure on the balconies were brought together in an elaborate fashion, giving the impression of a magnificent gateway. On the first levels of the ends of the bridge, on the Karaköy and Eminönü sides were shops. The floors where the shops were located were connected by two staircases to the platform at sea level. This platform, at the same time, served as a dock. The architecture of the shops and the structure of the staircases gave a Venetian air to the bridge.[100]

The 1902 proposal is also one of the rare projects whose fate is known. Said Pasha said in his memoirs, "The proposal was submitted at a time when Armenian uprisings in İstanbul were frequent. Therefore, for security reasons, the Sultan ordered that the design be worked on a bit more."[101]

In addition, in a response dated 9 March 1902 and sent by the Yıldız Palace Head Correspondence Bureau, both the rationale of Sultan Abdülhamid for rejecting Monsieur Pierre's project and a description of the kind of bridge he wanted are clearly spelled out.[102]

The construction of this bridge is both pertinent and apt, but bridges in famous European cities do not have shops and stores built on them. They have beautiful columns and the passageways for tramways, vehicles and pedestrians are separated from one another. Gas-fueled lanterns are located where needed. These enable people to come and go and allow them to take in the beautiful scenery on both sides. In addition, while the anchor points of bridges built over rivers are situated in sandy or swampy places, because the location where this bridge will be built is stable and rocky, the construction of the docks that will support it will be easier and sounder. Both sides of the bridge to be built should be stationary and the middle moving.

100 Çelik, *op cit.*, p. 74

101 İlter, *op cit.*, p. 82, *Said Paşa'nın Hatıratı*, Dersaadet 1327, v. II p. 212

102 BOA, İrade-i Hususi, 122, 29 Za 1319

25 - The Golden Horn Project of D'Aronco

26 - The French Proposal - 1902

The Golden Horn Bridge Project proposed by the French was shelved for this reason. The third proposal, which partially met the Sultan's description, came from D'Aronco. D'Aronco had nearly transformed the bridge, which had both passageways for tramway and horse-drawn carriages, and wide walkways, into a social project. He positioned eight, two-story balconies at both ends of the bridge; and the lower level, reached by stairs, was more functionally designed. The lower level would also serve as a dock for boats. It had another function; the seawater bath, extending under the bridge would be one of the constitutive components of the project. The bridge was decorated with lanterns fitting the sultan's description. Moreover, the people of İstanbul will be able to "take advantage of the beautiful scenery" by resting on banks placed on balconies at sea level.[103]

Following this proposal, Sultan Abdülhamid II approved a German Project, which would be used until it was damaged in a fire in 1992 and had to be pulled to shore. The project was submitted to the Sultan in autumn 1902 and signed in 1906. However, construction was unable to begin during Sultan Abdülhamid II's rule; in 1909, the Constitutional government concluded an agreement with the same German company for a bridge that would serve İstanbul for 83 years. The bridge, which is located today on the shore of Feshane, serves as the platform for various cultural activities.

103 National Palaces Painting Collection. No. 12/2800

Aurigue's Galata-Süleymaniye Suspension Bridge Project

The reason for keeping this plan separate from the heading of the Golden Horn Bridges is that will connect Süleymaniye and Galata, two districts not on the Golden Horn, to one another. The person who came up with this idea was Aurigue, Director of the Municipal Public Works Commission.

As stated above, Bouvard, who came to İstanbul in 1906 at the insistence of Sultan Abdülhamid II, declared that he would not be able to implement designs without first conducting various assessments and a city assay being made. For this purpose, Bouvard had Schrader draw up a city plan and left Aurigue in İstanbul to oversee public works. Aurigue, who, upon Bouvard's recommendation was appointed to the Municipal Public Works Commission, drew up a plan, which was published in French and Turkish, for the improvement of fire stations and public works in İstanbul.

Yenikapı-Unkapanı, Aksaray-Beyazıt streets were opened in this period; fire and sewerage projects were completed in Beyazıt and Aksaray. New neighborhoods and wide streets were built from Koska to Aksaray. Aurigue came up with solutions during his term to various problems of İstanbul, and developed certain tunnel projects. Among these was the Galata-Süleymaniye Bridge project.[104] What Aurigue had in mind was a transit way extending from the historical peninsula to Beyoğlu, spanning but not stopping at the Golden Horn. The suspension bridge, with one foot in Süleymaniye and the other near Galata Tower, would connect the two banks of the Golden Horn via a flat plane. This suspension bridge, which would be 1,200 meters in length, would add a new element to the basic silhouette of İstanbul seen from Galata. Most likely, the basic problem was this. Nowadays, the metal suspension bridge being built between Azapkapı and Unkapanı for the metro crossing is also an affront to the İstanbul silhouette. It's possible that the Ottoman ruler noticed this drawback once before and had, for this reason, affixed an annotation to the idea of a Galata-Süleymaniye Bridge. Unfortunately, apart from Prof. İhsan Mungan, the new design has not come under the fire of any academician. Any of the bridges built over the Golden Horn up till today have been made under the line of the silhouette and void of any unnecessary elevated parts. However, the towering buttresses and suspension cables of the bridge obscure the Bayezid Mosque, Fire Tower, Nuruosmaniye Mosque, Sultan Ahmet Mosque and Hagia Sophia when viewed from Azapkapı, and Süleymaniye and the Fatih, Yavuz Selim and Mihrimah Sultan mosques when viewed from the Golden Horn shores in Kasımpaşa. It's laudable that Ottoman administrations never made this mistake. But at its meeting held in Brazil, UNESCO objected to the design of the bridge in an effort to protect the natural fabric and historical appearance of İstanbul, which has World Heritage status. Rapporteurs carried out on-site inspections on the Golden Horn and they requested that the bridge's buttresses, which reached 50 meters, be lowered.[105]

The Golden Horn-Piyale Pasha Canal

Evliya Çelebi's Book of Travels describes an interesting initiative in the history of Ottoman housing policy and city planning that is not commonly known today. According to the famous traveler, in the face of the population growth witnessed in the second half of the 16th century, in order to reduce the population concentration in the historical peninsula, which was surrounded by the city walls, Sultan Süleyman the Magnificent ordered a number of state officials, among them Piyale Pasha, to build social complexes to serve as new settlement centers. Piyale Pasha received orders to build a shipyard where he had chosen for a prospective social complex. The connection between the shipyard and the sea, by a canal dug from the shore of the Golden Horn to the Piyale Pasha district, which is where the social complex is today.[106] Piri Pasha Stream was rehabilitated for this purpose. The canal in question would connect the social complex to be built with the center in Kasımpaşa; it would also link it to other parts of İstanbul through the Golden Horn, thus making this outlying district more suitable for settlement. The canal was made ready near the time Sinan's social complex was completed. Thus, the first canal project of the empire was executed. However, the new neighborhood created around the social complex did not last long; after its creator died, maintenance of the canal was neglected and alluvial deposits carried by flood waters filled the bed, making transportation difficult. Consequently, the neighborhood began to be abandoned and replaced by meadows and gardens.[107]

In 1569, approximately four years before a wide and deep canal project enabling heavy war ships built in the Dockyard to reach the Golden Horn was carried out, again during the time of Sultan Süleyman the Magnificent, the Ottoman armies, under the command of Mustafa Kaptan Pasha, were on the steps of Central Asia to connect the Don and Volga rivers to one another via canal.

104 Ergin, *op cit.*, p. 50

105 *Zaman*, 24 March 2011

106 Evliya Çelebi, *Evliya Çelebi Seyahatnamesi*, İstanbul: YKY, 2003, vol. 1, p. 101

107 Demiriz, Y. and B. Tanman, *Dünden Bugüne İstanbul Ansiklopedisi*, v. 6, p. 255

Canal Projects

The Golden Horn-Black Sea Canal Project

108 Demiriz Y.; Tanman B., *Dünden Bugüne İstanbul Ansiklopedisi*, v. 6, p. 255

The idea of a Golden Horn-Black Sea Canal that is mentioned among Ottoman projects came close to being an urban renewal project. Fundamental to the idea of connecting the Black Sea to the Golden horn through the Kağıthane Stream was industrial facilities planned for construction in Kağıthane. Bosphorus traffic in the middle of the 19th century was quite active. Due to the absence of large ships, transportation between the two continents was done by small boats and caïques. Conceivably, one of the benefits of developing project would be moving some of the traffic on the straits to the canal. The Black Sea-Marmara connection was principality a harbor, which was central to the Kağıthane project. At the same time, the project, the object of which was overseeing ships that would be diverted to the Marmara, called for the construction of a canal reaching a length of 31 km.[108]

Another project with similar aims was prepared 350 years earlier under Sultan Süleyman the Magnificent by Piyale Pasha. This project, too, had as its goal moving the heavy traffic on the Golden Horn to other centers.

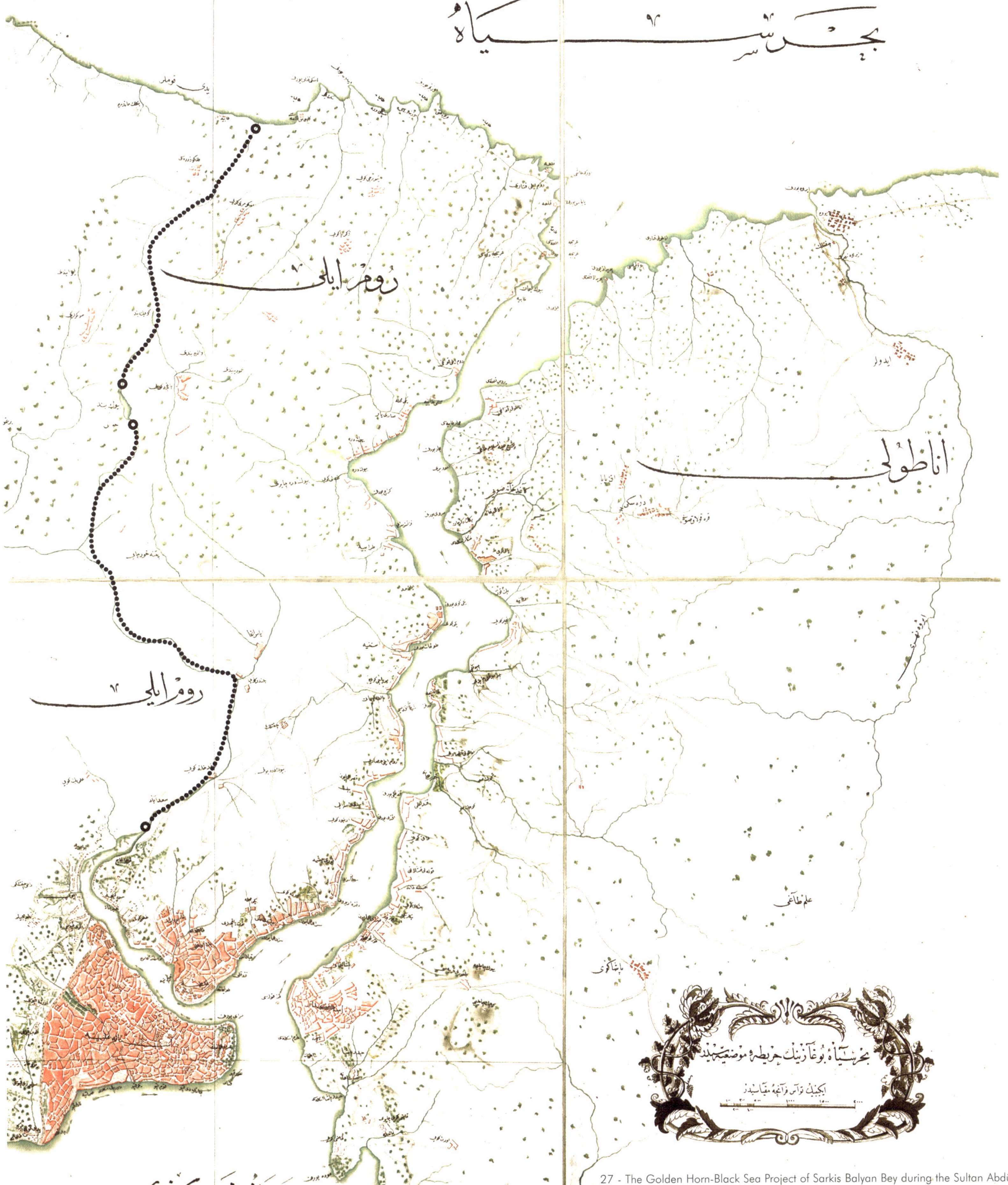

27 - The Golden Horn-Black Sea Project of Sarkis Balyan Bey during the Sultan Abdülaziz period

109 Çeçen, Kazım, *Osmanlı İmparatorluğu'nun Doruğu 16. Yüzyıl Teknolojisi*, İSKİ İstanbul 199, s.51

The Don - Volga Canal Project

With Ivan IV's conquering of the Kazan Khanate in 1552 and the Astrahan Khanate in 1554, Russia reached the shores of the Caspian Sea. Russia's new target was the Crimea Khanate. However, Crimea was under Ottoman patronage. The rulers of the Sublime Porte had to impede this expansionist policy of Russia. A canal between the Don and Volga rivers would make reaching the Caspian Sea possible and also serve as a barrier against Russia. By opening the canal, the Astrahan Khanate would once again be accessible and Russia would be prevented from entering the Black Sea. Moreover, the provision of requisite military supplies on campaigns to Iran and Central Asia could be done by sea.

At the beginning of 1569, the equipment needed to excavate the canal was made ready in İstanbul and shipped to the Azak Sea. In April, 3,000 Janissaries, under the command of Mustafa Kaptan, were dispatched to the area on the same route. Then in August of the same year, a second dispatch of Janissaries, totally 5,000, was made. In addition to Janissaries, 6,000 laborers, and 3,000 Sanjak soldiers and cavalrymen, engineers and specialists were sent. The Tatar Khanate contributed roughly 30,000 soldiers to this military force dispatched from İstanbul. According to Ottoman sources, one-third of the planned 7 km-long canal was dug. (While a length of 7 km may be mentioned in the sources, the closest distance between the two rivers today is 50 km.) But the Crimean Khanate united with the Russian Tsar and mobilized. This, along with the impact of severe winter conditions, led to failure on the part of Governor Kasım Çerkez Pasha. Russian Tsar Peter, who lived from 1672–1725, tried opening the same canal. This attempt, as well, came to naught. Construction of the canal would have to wait until 1952. Russia built a 45 km-long canal (the Volga-Don Canal) 16 km south of the point designated by Ottoman engineers, by creating a series of artificial lakes large enough to allow 5,000 ton-ships to pass through.[109]

110 Çeçen, *op cit.*, p. 52

111 For pertinent correspondence, see BOA, HAT, 1319/51446; BOA A.DVN. MHM.d 067; BOA, HRT 0542 (Çizim); BOA, İMMS, 9-386

Marmara - Sapanca - Sakarya Canal Project

Over the years that excavations of the Don-Volga Canal were being carried out by the Ottoman State, the architect (Mimar) Sinan and Engineer Kiriz Nikola were conducting expeditions for a canal project that would connect the Marmara Sea to the Sakarya River by passing through Sapanca Lake. According to Hammer and Purgstall, a 15 km-section was dug.[110]

The masts of the Ottoman naval fleet were procured from Walachia and Moldova, and the timber for its ships' hull and decks came from Kocaeli, Eskişehir and Bolu. The construction of a Sakarya-Sapanca-İzmit Gulf link was under consideration because transportation over land in those days was extremely difficult. Timber would be shipped from Sakarya to Sapanca Lake on ships and, if need be, war ships would be built on Sapanca Lake or in the İzmit Gulf.

Sultan Murad III sent a directive dated 21 January 1591 to the İznikmid (İzmit) and Sapanca Kadis. In it he commanded, "I wish for Sakarya River to flow into Sapanca Lake and Sapanca Lake into the İzmit Gulf. Have whatever is needed done, without negligence or slackness. Have the distance from Sakarya to the lake and to the gulf measured. Meanwhile, find out if there are mills, dairies, farms and the like there; if there are, if they can be relocated. Have comprehensive information prepared in writing and delivered urgently." Then a delegation of architects and engineers were sent to the region. Edicts were sent to the governors in Karaman, Sivas, Maraş and Erzurum to procure laborers for the construction of the canal. It was decided that there was a need for 30,000 of them. The job of excavation and supervision of the project was given to Sokulluzade Hasan Pasha. The former treasurer of Budin, Ahmed Efendi, was appointed Canal Director and Mustafa Efendi as secretary. Following the appointments, expropriation was carried out in the region and some of the farms there were relocated. In later periods, Mehmed IV, Mahmud I, Mustafa III, Mahmud II and Sultans Abdülmecid and Abdülaziz issued edicts for a project having the same ends to be implemented.[111]

Another canal project, about which Sultan Abdülaziz was adamant, was between Rosova and Constanta. Like the Russians, who began excavating between the Don and Volga rivers using prison camps established in the 1950s, during the same years, fervent work on another Ottoman project was begun on the western shore of the Black Sea.

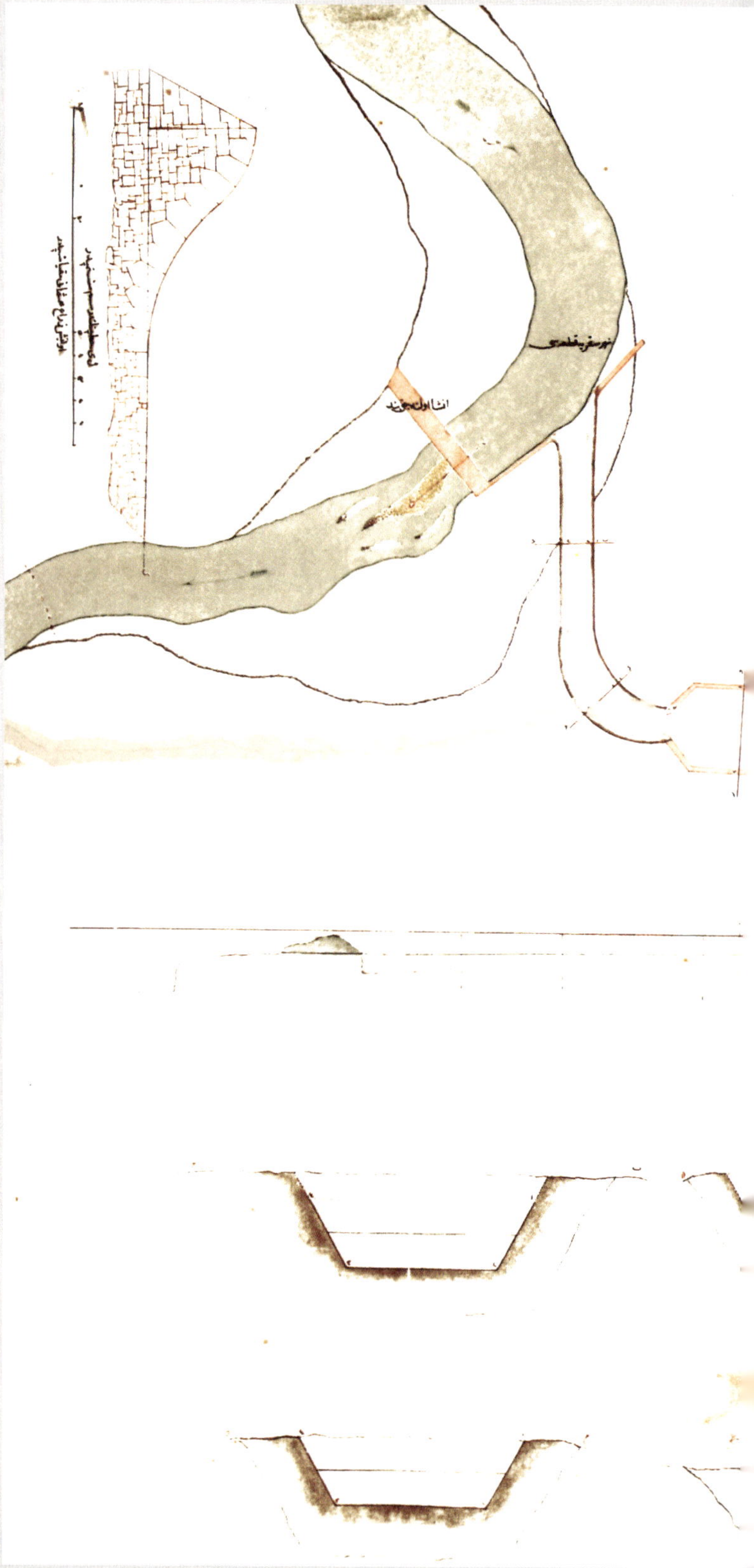

28 - Marmara - Sapanca - Sakarya Canal Project - 1923

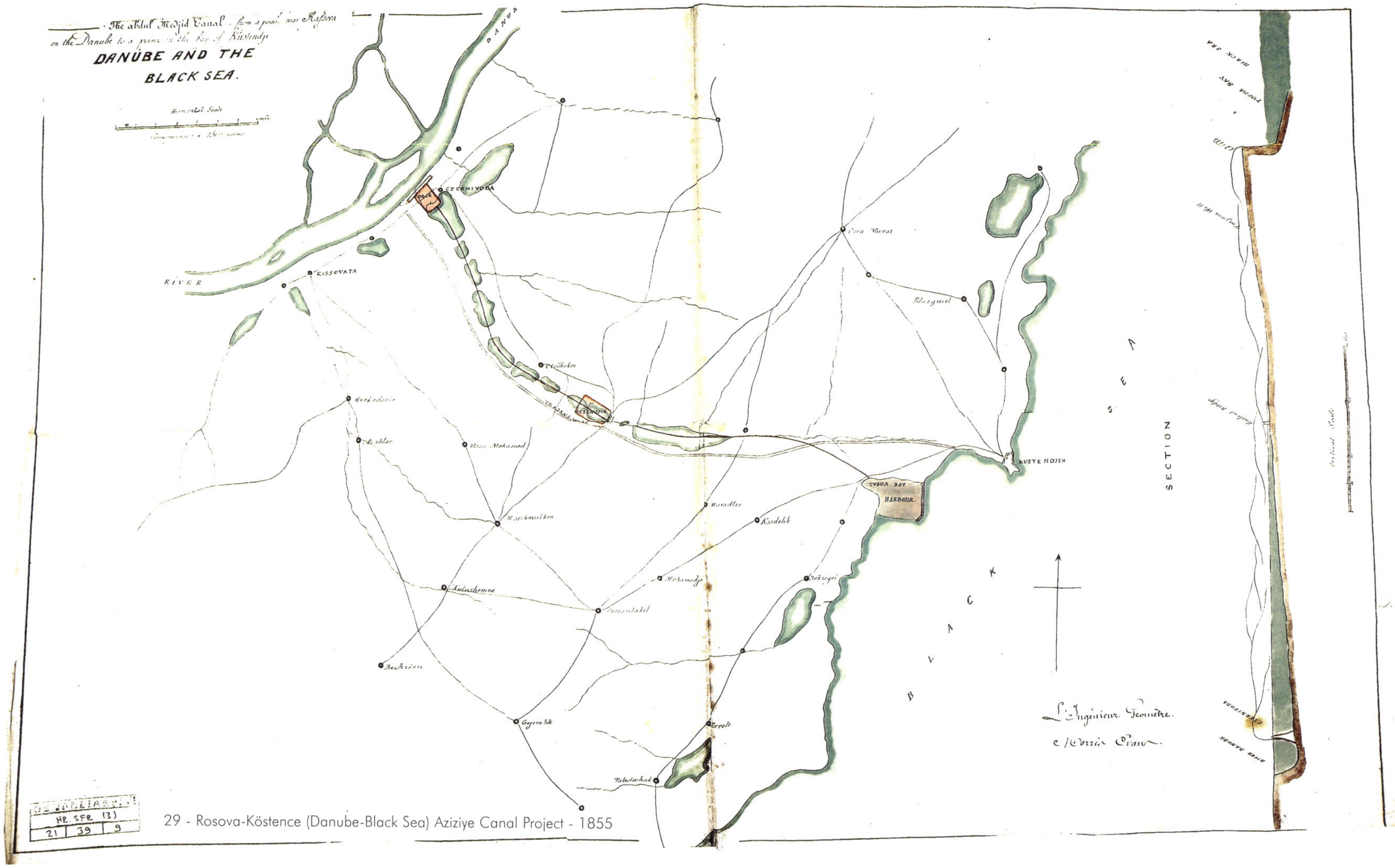

29 - Rosova-Köstence (Danube-Black Sea) Aziziye Canal Project - 1855

Rosova-Constanta (Danube-Black Sea) Mecidiye Canal Project

The first serious efforts to open a canal go back to the 1830s, following the Treaty of Edirne. Through this treaty, Greece acquired independence and special status was given to Serbia, Wallachia and Moldavia. The Ottoman State had partially lost its position in the Balkans. With the new status won by the Wallachia, Moldavia and Danube principalities, they had become a threat to the Ottoman commercial monopoly in the region. Nevertheless, important port towns remained under the sovereignty of the Ottoman State, which protected the existence of commerce in the Balkans. Constanta was just one of them. While control of the delta where the Danube River flowed into the Black Sea was in the hands of the Ottomans, a delta is far from being a port. Therefore, if Constanta Port, the closest, could be linked to the Danube River, which has sufficient capacity, shipping done as far as Austria would end at a port and reach the Black Sea.

Naturally, the project was being viewed favorably in Austria, but this initiative would leave the Crimea Port, which was one of the largest ports on the Black Sea, inoperative. And Russia would not be in the interests of Russia. Getting wind of the move and calculating the loss in taxes it would experience, it began lobbying against the project.[112]

112 Spulber, Nicolas, "The Danube – Black Sea Canal and the Russian Control over the Danube," *Economic Geography*, July 1954, issue 30, pp. 236–245

Throughout Mahmud II's period, Russia maintained its counter attack and the project was postponed numerous times. But Austria and Austrian entrepreneurs did not intend to give up pursuing the project. In fact, in 1839, the Austrian Szechenyi Company was awarded the right to build a railroad parallel to the canal the Ottoman government was thinking of building between the Port of Constanta and Rosova. However, bribery and threats were factors in this that prevented the construction of a railroad during this time.[113]

The Crimean War (1854–1856) added a strategic dimension to the project. Now, along with the Ottoman State squaring off against Russia, there was England and France, as well. This project became the most practical way to get military contingents and logistic support on the Danube front to the Black Sea. In the summer of 1855, work started by the French Engineer Charles LaLanne was completed by the end of the year and Rosova was connected to the Port of Constanta by highway. The construction of the road did not eliminate the need for a canal. In the meantime, the Austrian company that had done the initial work got involved again and renewed its efforts. On 23 July 1855, Sultan Abdülmecid gave the job of constructing a port in Constanta, a canal between Rosova and Constanta, and a railroad on the same route to a company represented by Forbes Campbell, which would be established as an Austrian-English-French consortium.[114]

With the signing on 30 March 1856 of the Treaty of Paris, which ended the Crimean War, the course of the project, as well, changed. According to Article 7 of the treaty, unrestricted transit on the Danube would be reestablished and, to maintain it, the Danube Commission would be created by the representatives of the states that were signatories to this treaty. The consortium, consisting of England, France, Austria, Russia, the Ottoman State, the Kingdom of Prussia and the Kingdom of Sardinia, made a number of decisions and took certain measures to clean the Danube delta. The existence of the consortium made Austria lose interest in the project. Therefore, on 25 May 1856, the Ottoman State relaunched the necessary efforts to get a railroad built between Rosova and Constanta, which had previously been linked by highway. The laying of 57 km of railroad was completed on 4 October 1860 and put into service with a ceremony.[115]

The existence of the railroad line, along with impact of difficult conditions, led Ottoman rulers to postpone the construction of a canal. At

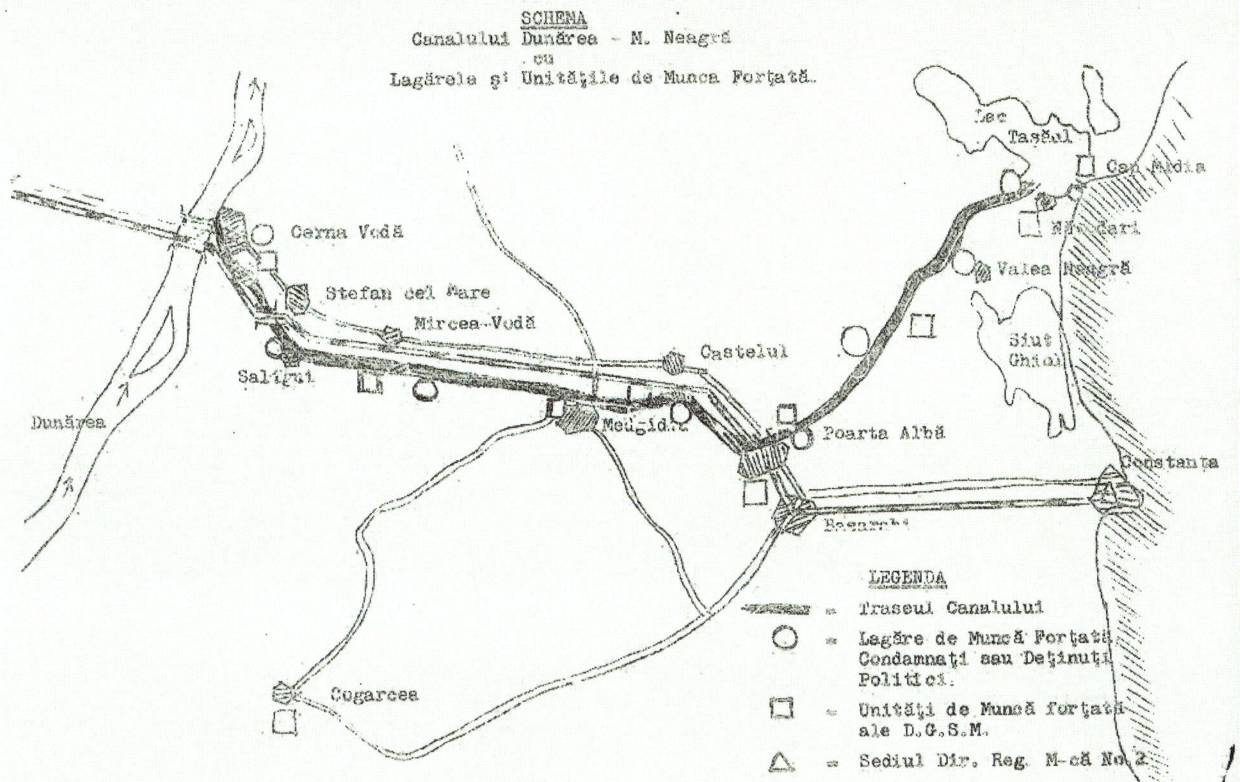

30 - Rosova-Constanta, Romania - 1855

the beginning of the 20th century, there was a new and independent state: Romania. The project, which would acquire national importance, was put off until the 1950s due to the Balkan Wars, the First World War, which immediately followed, the Great Depression in the 1930s and the Second World War.

Romania, which broke off from the Ottoman State as a kingdom, had a change in regime in the 1950s. In 1947, the Communist People's Party was at the helm. During the same years, the Soviet Union began excavations on another Ottoman canal project. Not only was a regime transferred to the other end of the Black Sea, but also, along with this regime, the idea of implementing Ottoman projects that were labor intensive. The new regime placed thousands of Romanian opponents into large camps. The project, which is remembered with sadness despite its economic usefulness today, was constructed by the heart-wrenching stories of thousands of Romanians who were imprisoned in these concentration camps. It is not known how many people died during the building of the 67 km-canal, on which some sources say 100,000 prisoners worked.[116]

Inadequate nutrition and health conditions, inadequate protection against severe cold, and diseases such as tuberculosis caused the deaths of tens of thousands of Romanians who worked on what is now referred to as the "Death Canal."[117]

113 Petrescu, Florian, *Din istoria dacoromanilor si a altor popoare,* "Romanii si primele proiecte de construire a canalului Dunare-Marea Neagra (1838-1856)", 2009, Verus, pp.136--139

114 BOA, HR.SFR.3, File no: 21, Jacket no:17

115 Petrescu, *op cit.*, p. 141

116 *Time*, August 24, 1953

117 Tismaneanu, Vladimir, *Stalinism for All Seasons: A Political History of Romanian Communism.* California University Press, 2003, p. 139

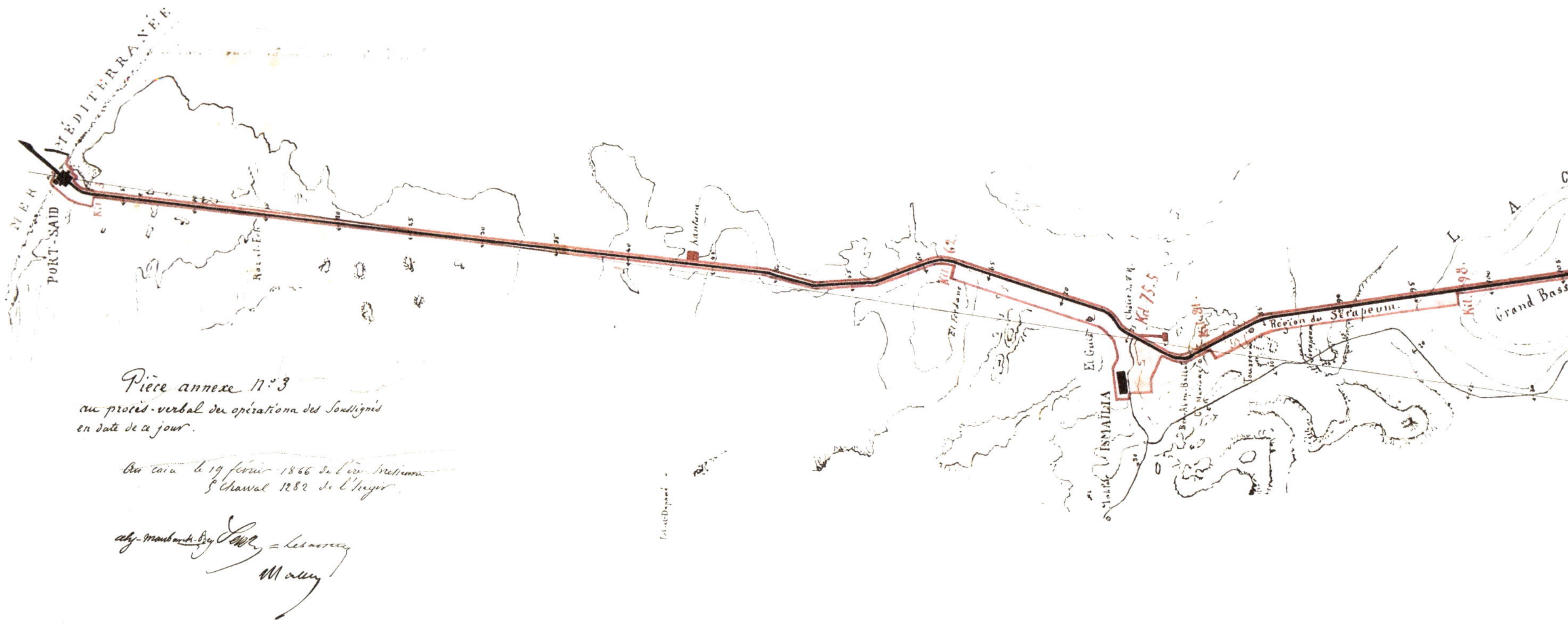

Red Sea-Mediterranean (Suez) Canal Project

Prior to the opening of the Suez Canal, the Suez isthmus geographically linked the continents of Asia and Africa in the ancient world; it also separated the Red Sea and the Mediterranean, and therefore the Indian and Atlantic Oceans from one another.

In the early eras when maritime navigation wasn't developed enough to compete with overland shipping, the Suez was a communications bridge between continents and had major geopolitical importance. Therefore, Suez was the scene for many military political and economic events. Its geographical location made it a perfect crossing point for invasions: using this international bridge, the Egyptians swept into Asia, the Arabs made major inroads into North Africa the 7th century, and finally, the Turks reached the shores of the Red Sea.

In later centuries, development of maritime navigation and new discoveries changed the direction of commercial roads. At the end of the 15th century, the Portuguese went around the Horn of Africa and reached India. The ports in Suez and the Nile Delta lost their importance. However, this was an onerous solution. In 1830, it took ships using the Cape of Good route between England and India 7 months to complete the journey. Because the route to India was so long, the idea of the need for building a canal through Suez was put forward.

Turks, who at the beginning of the 16th century dominated the region, had a fleet in the Red Sea as well as the Mediterranean. They even had a large dockyard at the Red Sea. Commanders like Sinan Pasha and famous sailors like Piri Reis and Murad Reis went on many campaigns with this fleet. The traveling of the Mediterranean fleet when needed to the Red Sea had an

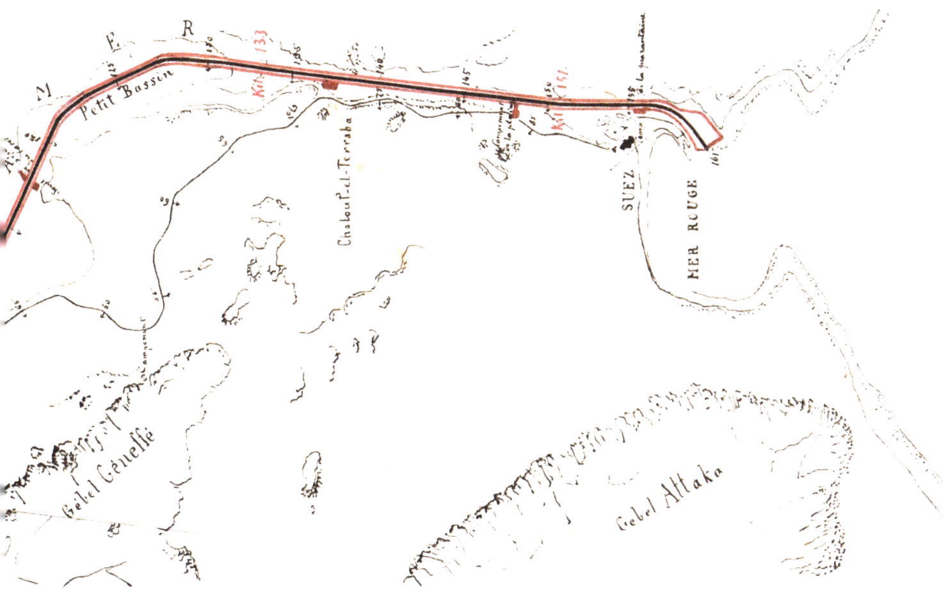

31 - Red Sea - Mediterranean (Suez) Canal Project - 1866

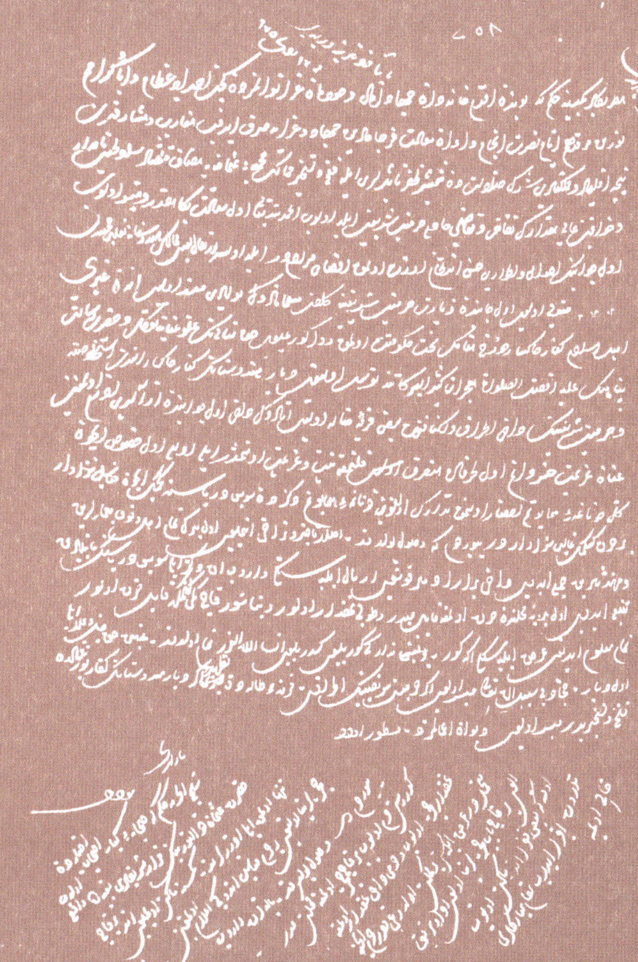

32 - Sultan Selim II's edict to the governors in Egypt - 1568

impact on the Indian Ocean, which had grown in importance during those years through this route. In order to for the state be able to completely control Arabia, Sudan and Ethiopia, opening the Suez Canal was imperative.

The edict Sultan Selim II sent to the Governor of Egypt in 1568 was the first step in this direction. In it, it was indicated that the Portuguese harassed Indian Muslims and that they blocked those wanting to reach Mecca for Hajj using the Indian route. It was emphasized that to save India from Portuguese assault, a canal had to be constructed in the Mediterranean so that the fleet could reach the Red Sea.

Architects and engineers were sent for this purpose to measure the distance between the Mediterranean and Suez. Sultan Selim II want to find out whether or not it would be possible to construct a canal, how long it would be and how many ships would be able to pass through it.[118]

118 BOA,A. DVNS.MHM.d, 7/721

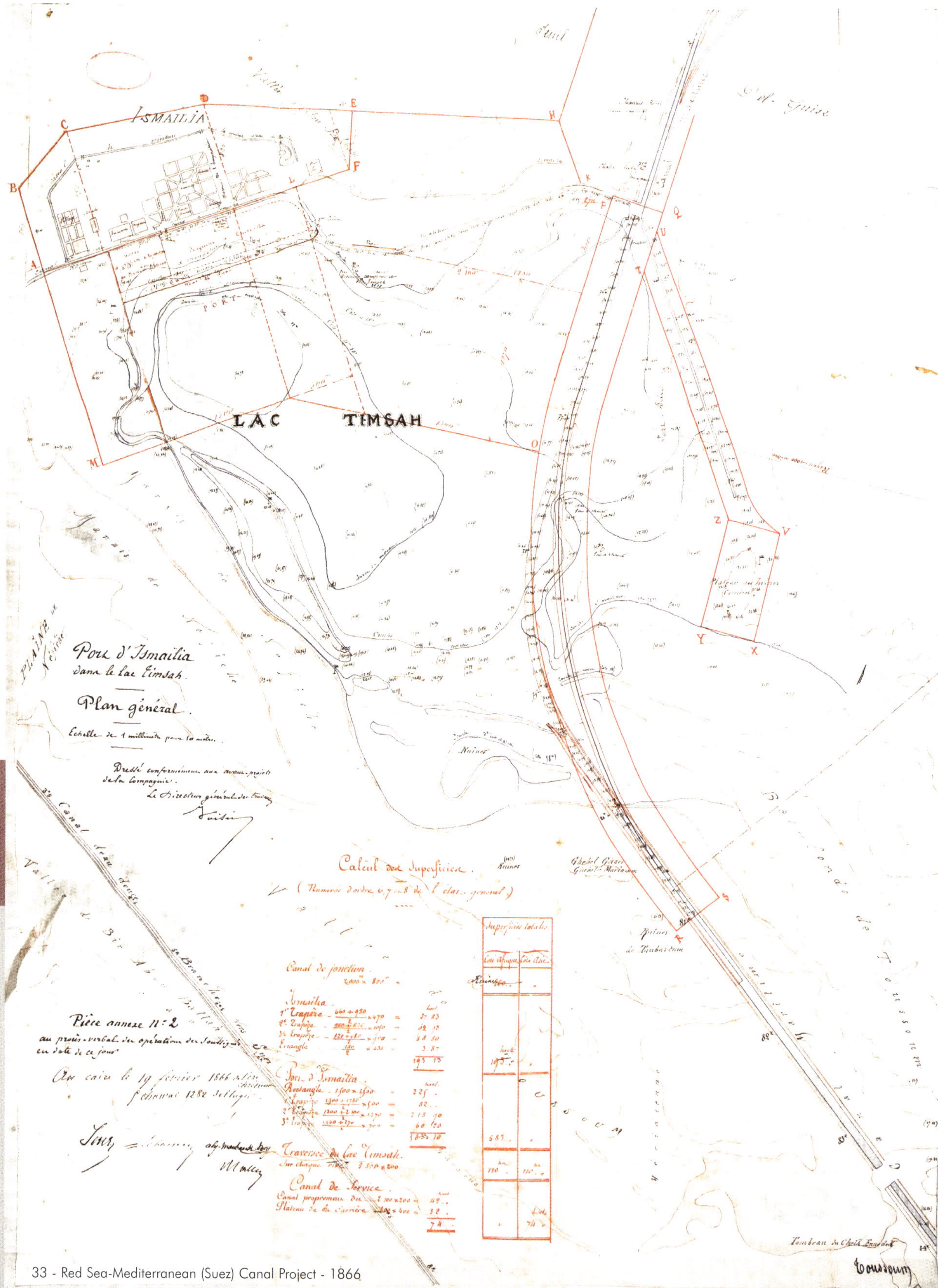

33 - Red Sea-Mediterranean (Suez) Canal Project - 1866

Kara İbrahim Pasha, who served as Governor of Egypt between 1669 and 1673, had a different canal idea. He thought that instead of a canal between the Red Sea and the Mediterranean, one could be built between the Nile River and the Red Sea, and in that way the Mediterranean could be reached. But Egyptian notables were afraid that with the flow of the Nile into the Red Sea, there would be a reduction in non-irrigatable land, so the Pasha abandoned this initiative.[119]

Admiral- in-Chief Ali Uluç proposed a Mediterranean-Red Sea Project to Sultan Mustafa III; supporting Uluç's idea, the Sultan had Baron de Tott carry out the necessary analyses. Nevertheless, the idea could not be realized.[120]

Lewis writes that Emir Mehmed bin Emir Hasan, author of *Hadis-i Nev*, encouraged Murad III to open the Suez Canal in order to strengthen the state vis-à-vis the threat to Muslim countries presented by Europeans settling on the shores of America, India and the Gulf of Iran, and, to stay abreast of the threat posed to Muslim commerce by Europeans.[121]

The idea of opening a canal was forgotten for a long time, only to reappear during Napoleon's Egyptian campaign. The engineer Le Pere joined Napoleon's army in 1798 as the engineer of the campaign delegation. He was assigned by Napoleon the job of drawing a map of the route of the ancient canal that had been opened by the Pharaohs. Napoleon, viewing the Indian route as a problem, took great personal interest in the matter and traveled kilometers on horseback to inspect work on the canal. He gave Le Pere this order:

"This problem is very big. Issue this memorandum. And put pressure on the Turkish government to follow up this tremendous and beneficial project."[122]

Furthermore, in subsequent years, during the time of Sultan Mahmud II, we know that a canal, called the Mahmudiye Canal, was opened from the Nile

River to Alexandria, that it was for irrigational purposes, and that the canal was cleaned under the governorship of Said Pasha.[123]

Finally, on 17 November 1869, three engineers, a Frenchman, an Englishman and a German, conducted a joint study, the results of which showed that there was no difference in level, just as Le Pere had calculated. The carrying out of the canal project was left to Ferdinand de Lesseps, who taking advantage of the changing political climate in Egypt, succeeded in overcoming many problems and diplomatic obstacles. In spite of everything, the project became a reality with Ottoman consent in the form of an edict by Sultan Abdülaziz dated 19 March 1866. Sultan Abdülaziz not only approved the agreements concluded between the Canal Company set up to build the canal and Said and İsmail Pashas, but also gave state assurances to the project.[124] Hence, the first Ottoman canal project was actualized. At the same time, İsmail Pasha established contact with Frederic Auguste Bartholdi to built a monument, which due to its size, would spark interest at the opening of the canal.

119 Öztuna, Yılmaz, *Osmanlı Devleti Tarihi-2-*, Kültür Bakanlığı Yay., V. II, Ankara, 1998, p. 364

120 Uzunçarşılı, İ.Hakkı, *Osmanlı Tarihi*, 3. Bsk., TTK., Ankara, 1983, v. III, p. 90

121 Bediz, Danyal, "Süveyş Kanalı'nın Ehemmiyeti," *İktisadi Uyanış*, p. 49, April 1952, p.19

122 Oral, Cavid, *Akdeniz Meselesi*, Bugün Matbaası, Adana, 1943, p. 191

123 Ateş, Toktamış, *Siyasal Tarih -I*, İstanbul Üniversitesi İktisat Fakültesi, İstanbul, 1989, p. 305

124 Berkol, *Süveyş Kanalının Hukuki Statüsü*, Alaeddin Kıral Basımevi, Ankara, 1941, p. 6

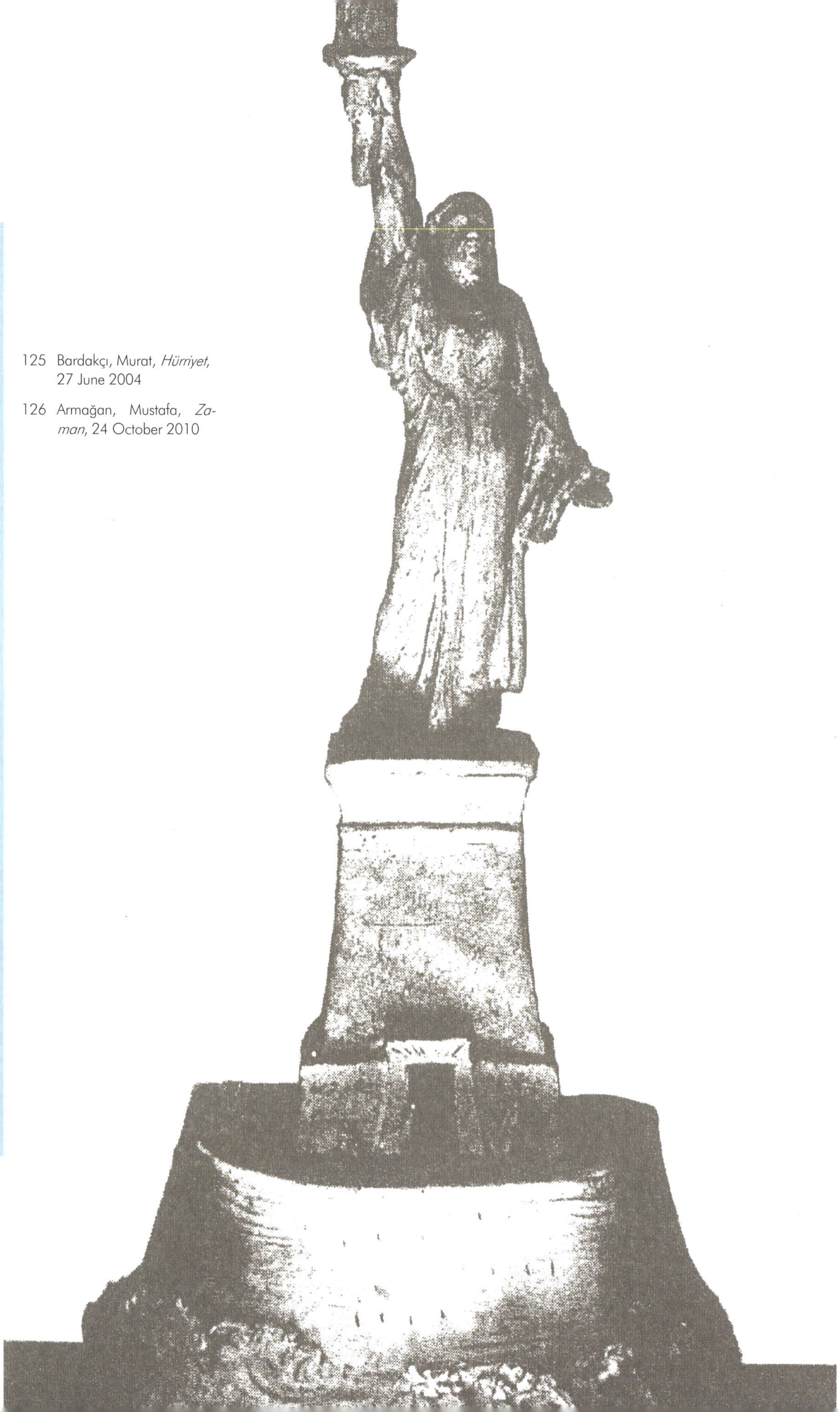

125 Bardakçı, Murat, *Hürriyet*, 27 June 2004

126 Armağan, Mustafa, *Zaman*, 24 October 2010

Light Emanating from Asia or the Statue of Liberty

This heading reflects the subject of a polemic between two researchers that became a topic of discussion in Turkey in 2004 and really confused readers. A journalist, providing references, wrote that the Statue of Liberty in the US had actually been ordered for the opening of the Suez Canal, and that a portion of its cost was paid as an advance by the Ottoman Sultan at that time, Abdülaziz Han. He said that after this payment was made, the statue was prepared in parts and loaded onto a ship in the Port of Marseilles waiting to depart for Egypt. However, when İsmail Pasha replaced Said Pasha as governor of Egypt, the former is quoted as saying "A statue like that in a Muslim country will create displeasure among the people," and, as a result, it was left in a warehouse in Paris to corrode.[125] Another Ottoman researcher/writer, however, retorted that the statue at no time had reached the production stage and argued that the Ottoman State during that era was not in a position to support other similar large projects.[126] The writer claims that while Frederic Auguste Bartholdi had prepared a project for the opening of the Suez Canal, even he denied that it bore any resemblance to the Statue of Liberty, referring to page 14 of Klaus Kreiser's article in Mukarnas journal.

Leaving debates aside, if the information and relevant visuals in the article mentioned are taken into consideration, we believe that the project designed for Egypt, despite claims by the designer of "chance resemblance," is very similar to the main contours of the Statue of Liberty. There is no mention in the article that this prototype of the statue now in the US would be built on a smaller scale. We also do not know the dimensions of the project submitted or whether Abdülaziz Han, "the first Ottoman Sultan to be made into a statue," even had any relationship at all with the matter. The project was unable to move beyond the prototype stage; what's more, Bartholdi later gave presentations on Mehmed Ali Pasha to İsmail Pasha.

Frederic Auguste Bartholdi made a name for himself at the Paris Exposition Universelle with the bronze statue he made of General Jean Rapp in 1856. Following the exposition, he went with the painter Leon Gerome to Yemen and Egypt. After the Egypt trip, he made a statue of the famous philologist Jean Francois Champollion; the plaster statue was exhibited in the Egypt stand at the 1867 Paris exposition. Bartholdi met İsmail Pasha when he was in Egypt, and learned that the pasha was going to have a monumental work made for the opening of the Suez Canal. When he returned to France, he prepared a prototype.

"The statue will be in the form of a village woman wrapped in clothing of the time of the Pharoahs, and in her hand she will hold a torch symbolizing that the light of Asia came from Egypt"... we think that this is how the description of the order was made to Bertholdi.

The same year, İsmail Pasha visits Paris; getting wind of this, Bartholdi immediately contacts the pasha. Bartholdi has applied to the governor to get the project implemented. Now, here is where it gets interesting: İsmail Pasha does not like Bertholdi's design. Yes, the only reason the project, whose resemblance to the Statue of Liberty was described as "chance," was not ordered for the opening of the Suez Canal is that Bartholdi was unable to get the pasha to like it.[127]

The final note we will make in this section is that the prototype designed for Egypt is what Klaus Kreiser photographed at the Barholdi Municipal Museum in Colmar and whose photograph he used in his article.

127 Kreiser, Klaus, *Mugarnas*, Issue 14, pp. 104–108

Sultan Abdülhamid Lake and Dead Sea-Mediterranean Canal Project

128 BOA, Y.Prk.Tkm, 34/58 29.za.1312

This project was noted for the first time in a work entitled Sultan Abdülhamid's Political Memoirs under the heading "Competition with the Suez Canal (1898)."

As it is known, on 11 June 1882 in Alexandria, an Egyptian driver and his European customer fought over a disagreement regarding fare. The Europeans in the city withdrew into their homes and began firing on the local people. Moreover, the English fleet started pelting the city with cannons. The English invaded from the sea and, taking Arabi Pasha captive in the Al-Kabir command post, officially occupied Egypt.

England declared that the occupation would not last long and as soon as order had been restored, it would withdraw its army from Egypt. It even signed an agreement in 1887 stating that it would leave Egypt within three years. However, their promise was never kept. With the occupation of Sudan in 1898, England settled in Egypt permanently.

In one sense, the heading, "Competition with the Suez Canal" is the official recognition of Sultan Abdülhamid after much lengthy endeavor. Sultan Abdülhamid II, who leaned in the direction of alternative projects after this date, was making an effort to weaken the impact of the Suez Canal on the region.

As a matter of fact, earlier, a project[128] had been put on the table on 24 May 1895, as a kind of exercise in idea generation, to "extend the railroad

34 - Sultan Abdülhamid Lake and Aqaba - Mediterranean Canal Project

line running from İstanbul to Baghdad as far as Basra, thereby creating a railroad line that served the same function as the Suez Canal."

Under this heading in his memoirs, Sultan Abdülhamid had this to say about the project: "A retired Prussian military officer submitted a proposal to construct a canal that could compete with the Suez Canal. According to what he said, research done in Wadi Araba (Araba Valley) showed that the Dead Sea and the Aqaba Gulf used to be connected. If a roughly 90 km-area were excavated here, it would remain 25–30 km underwater. This would create a large lake in Sharia-al kabira (Jordan) and the Al-Ghawr valleys; life and commerce would, just as they had in the past, would develop on the shores.

By assaying the topography of this region, it will also be useful to find a way of connecting this lake, which will bear my name, with the Mediterranean. The canals thereby created will be as important politically as economically. Part of the traffic will be transferred to the Arab-Syria route. Turkey and other states that are friends of Turkey will no longer be dependent upon the Suez Canal.

I believe that it will be worth seriously examining the plans of the Prussian officer. It will be appropriate for a special commission that will be able to review this project in detail to be set up."[129]

As can be seen, the approach taken by Ottoman sultans to projects similar to this one is extremely clear: immediately setting up a commission and

129 Sultan Abdülhamid, *Siyasi Hatıralarım*, Dergah, İstanbul, 1999, pp. 103–104

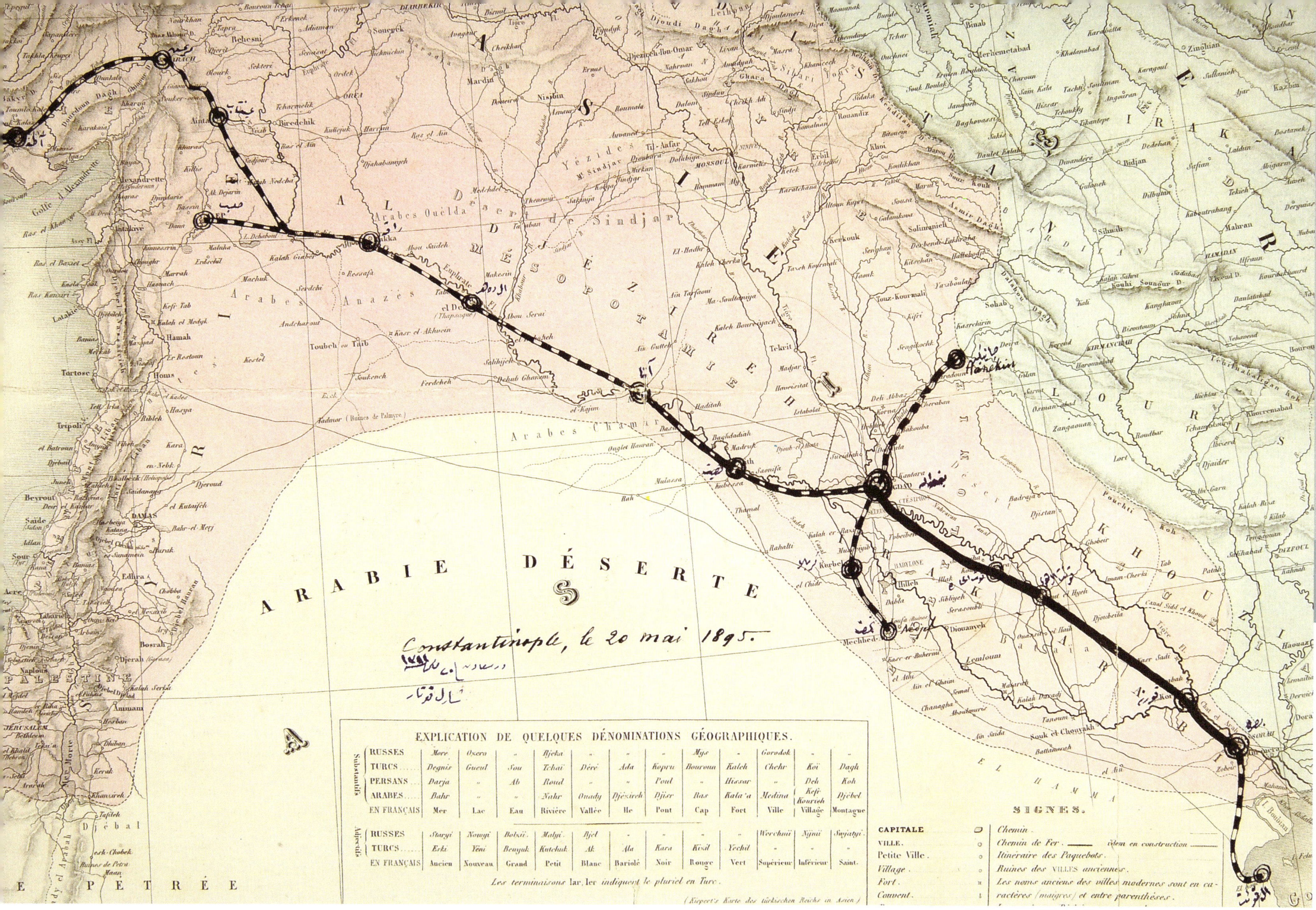

35 - Baghdad - Basra Railroad Project - 1895

the ordering a follow up of the matter; at the same time, making additions to the project by themselves. What was being proposed was a lake that would have agricultural and commercial impacts on the environment. Sultan Abdülhamid, realizing that by connecting Abdülhamid Lake, which was being merged with the Dead Sea, to the Mediterranean, there would also be benefits and political advantages, he stressed that studies for the canal needed to be conducted.

The Prussian officer corroborates the information about the region found in encyclopedias today. The Dead Sea is located in an active seismic region; in other words, in an earthquake zone. The floor of the Dead Sea is within a tectonic depression called the Rift Valley. This depression extends from Tabaria Lake, in the north, and, running along the Jordan River, to the middle of Araba Valley in the south; it's in a 300 km valley. This "depression valley" is seen as a

natural extension of the Red Sea. From available data, it can be said that Abdülhamid Lake, forming with the water coming from the canal to be dug from the Aqaba Gulf going into the depression, would be 72 km in length and 16 km in width. However, the existence of this depression was not only noticed by the Ottoman State but also by England. Initially not looking favorable on the construction of the Suez Canal, England later established a company for this purpose. Holding nearly all of its shares, its plans were based on alternatives routes to the Suez Canal not being developed. The Prussian officer's plan would not be forgotten in subsequent years, but when drawing a political map while taking this depression into consideration, the canal line he envisaged coincides with what is the Jordan-Israel border, which would make implementing the project impossible today.

Efforts have been made to overcome this impossibility. With the rapid shrinking of the Dead Sea, in 1973, the project was once again put on the agenda. But Jordan and Israel couldn't agree on the route of the canal, so it was shelved. The threat of extinction of the lake, whose name appears in the holy books of the three Abrahamic religions, has made the idea of linking the Mediterranean and the Red Sea a necessity. In 2005, the World Bank certified eleven companies to conduct a feasibility study; but it doesn't appear that disagreements will disappear. Jordan insists on the lake flowing through a Red Sea canal while Israel insists on it going through a Mediterranean canal.

Meanwhile, it doesn't appear that the new project includes the Prussian officer's idea of expanding the lake and filling in the large depression. The new project includes having sea water flow from the Red Sea or Mediterranean to The Dead Sea through underwater canals, taking advantage of the difference in height, generating electricity from the natural flow of water, and distilling drinking water from sea water through facilities to be set up for that purpose.[130]

If disagreements can be overcome, for the first 20 years, 3.3 million m^3 of sea water will flow into the lake, 1.3 million m^3 of drinking water will be obtained and 2,500 MW of electricity will be generated, thus another Ottoman project will have been implemented.

130 Gökçe, Deniz, *Akşam*, 26.10.2007

مرمره دكزى
قزل اطه لر
قنالى اطه
هكبه لى اطه
بيوك اطه
بورغاز اطه سى
صدف اطه سى
ديل

Sarkis Balyan's Heybeliada-Büyükada Bridge Project

For long time, the Islands were left to religious pilgrimage sites, fishing villages and summer homes due to difficulties in both maintaining law and order, as well as providing transportation. However, around the 1850s, with the start of steamship service, they became more valuable and Muslim settlement there quickened.

This rapid expansion in population created a number of problems. The first of these was communication between the islands and the mainland, as well as between each of the islands themselves. Increasing the number of trips made by the Şirket-i Hayriye (literally "The Goodwill Company"; a maritime transportation company) during this time, important steps were taken to solve the problem. However, a practical way of providing easy crossings back and forth between the two largest islands—Heybeliada and Büyükada—in particular, was necessary. This was because transportation between them was being provided by small caïques and the Şirket-i Hayriye boats, which came at specific times. Therefore, an engineer who was at the same time a resident of one of the islands, Sarkis Balyan, proposed to Sultan Abdülaziz, during the construction of Dolmabahçe Palace, to solve this problem. The proposal was a 1200 m-long suspension bridge. According to Sarkis Balyan's project, the suspension bridge would be 5–5.5 meters wide, and 1 kurus would be charged for using the bridge. Estimating that 300 persons would use the bridge every day, it would pay for itself in 50 years.[131] But the project never got off the ground.

Most probably the largest factor in the shelving of the project was that the shores connecting the two bridges to one another were 1200 meters away from each other. So even if the limits of construction technology of those years were stretched, exceeding 1200 meters for a suspension bridge would have been extremely difficult.

131 Kayra, *op cit.*, p. 66

Münif Pasha's Grand Ottoman Park Project

Münif Pasha was a statesman, translator and journalist who made important contributions to educational political during the Abdülhamid II era. He made many significant determinations with respect to İstanbul and came up with many important proposals. These include constructing wide streets and squares; illuminating streets to make it easier to work at night; numbering streets, residences and shops to prevent confusion and to give accurate directions; and a main thoroughfare that would extend from the Karaköy Bridge to Bab-ı Ali (the Sublime Porte) and from there to Topkapı via Aksaray. Münif Pasha mostly made these proposals in the *Mecmua-i Fünün*, the magazine he published. Another of Münif Pasha's dreams for İstanbul was the Ottoman State Park project. Münif Pasha's project was ambitious, but it is interesting that it was never carried out anywhere.

The park was to have been established on a rectangular area 3,000 meters in width and would have contained soil from all of the lands of the Ottoman State in Europe, Asia and Africa. In order to demonstrate which foreign states were adjacent to the Ottoman State, an amount of soil from those states would be displayed in the park. Similarly, lines would be drawn inside to be able to readily distinguish the regions from one another. Continents, seas, rivers, mountains, forests, mines, cities, roads, and telegraph lines, in short, all natural formations, to scale to the extent possible.

Seas and lakes were to be represented by pools filled with seawater. Islands with realistic features, including shores with ships anchored along them. Streams and rivers, too, would be added, making it look as though fresh water would be flowing through them naturally.

"Until now, nowhere else in the world has such a map ever been made. With its production in İstanbul, this beautiful work of art, pro-

duced with the help of the glorious sultan, will illustrate the wonderful convergence of states and geographical communities within the illustrious European states. It will show foreign travelers coming to İstanbul that it is something special worth travelling for and will serve to enhance the glory of the sultanate that much more."

Immediately under the proposal, Münif Pasha opened a section entitled "Costs and Revenues," where he laid out his ideas about the financing of the project. In his opinion, its organization and construction would cost five to six thousand pieces of gold, with the needed property being gotten from the state treasury. Münif Pasha even thought of the costs of "proper protection" of the park, for which he said seven to eight kurus would be needed. He also stressed in this section that the money could also be gotten by charging an appropriate entrance fee.[132]

In 2003, 120 works were chosen by the advisory committee for the purpose of being Turkey's showcase. Miniatures of them were made and a park by the name of "Miniaturk" was opened in Hasköy, İstanbul where they were exhibited outdoors. While it's clear that Münif Pasha's project was greater in scope, it can be said that "Miniaturk" is an Ottoman project.

132 Budak, Ali, *1453 İstanbul ve Sanat Dergisi*, 2010/7, pp. 73–83

37 - D'Aronco's Turkish-Greek War Victory Memorial Project - 1897

D'Aronco's Turkish-Greek War Victory Monument Project

D'Aronco was the Italian architect who had been invited in 1893 for the Ziraat ve Sanayi Sergi-i Umumisi (İstanbul Agricultural and Industrial Exposition), which was canceled due to the earthquake in 1894. Nonetheless, he stayed in İstanbul upon the request of Sultan Abdülhamid II and subsequently worked on many important projects. He left İstanbul in 1909 after the Sultan was disposed. Among the public structures D'Aronco had a hand in creating in İstanbul are the Etfal Hospital Clock Tower, the Haydarpaşa Military Hospital, the Imperial School of Medicine, along with Vallaury, the Nişantaşı Police Station, the Karaköy Mescid, the Janissary Museum, the Sultan Abdülhamid Fountains, the Ministry of Agriculture, Forests and Mines buildings, Şale Kiosk, Yıldız Porcelain Factory, the Imperial Porcelain Factory, Yıldız Palace Theater and the Sheikh Zafir Tekke (Dervish Lodge).

While he lived in İstanbul, between 1893 and 1909, D'Aronco witnessed many important incidents. Without a doubt, one of these was the Greco-Turkish War (1897). The war, the first and only one that Abdülhamid II entered on his own accord, concluded with the victory of the Ottoman Army.

In the aftermath of the Greco-Turkish War, having been thirsty for victory, the Ottomans were rightfully proud and exuberant over their triumph. Moreover, they had regained their esteem, which had been shaken in the 1893 War, thereby erasing the "Sick Man" image in the eyes of Europe. It was understood that heinous plans taken up against the state would not easily succeed. Moreover, throughout the Muslim world, from Egypt to India, celebrations were held, thereby reviving the idea of once against coming under Ottoman protection. This triumph regerminated the devotion felt toward the Ottoman caliphate.[133]

The Ottoman government called upon the Italian architect to build something that would be a permanent reminder of the victory. The

38 - Model of the project submitted to Sultan Abdülhamid - 1897

133 Hülagü, Metin, *Osmanlı Yunan Harbi - Abdülhamid'in Zaferi*, Yitik Hazine, İstanbul 2008, pp. 267–277

39 - Hicaz Railroad Telegraph Line Memorial Project - 1905

work was to embody the joy winning the war had brought to the entire Muslim world.

It was against this backdrop that D'Aronco designed the last great Ottoman triumphal monument, which would be held in awe by anyone seeing its immense silhouette, nearly 25 meters high. The large dome positioned upon its spherical frame was to create the impression of a "Hallowed Place." Its core, formed by massive buttresses, and four fountains, on the outer frame, typologically resembles a mosque. However, the guns and ensigns adorning it add a memorial-like flavor to the project. Four, Egyptian obelisk-like columns, each with six steps leading up to them, would stretch upwards from the corners frame. Each of the four sides supported by the columns is to have similar design so that the same feelings are elicited regardless of the direction from which the monument is viewed. The Ottoman

crest on the doors gives it the appearance of a Triumphal Arch. Clocks that look like medallions are placed between the fountains and the columns. Around the clocks are strung Ottoman pennants.[134]

Before the jubilance had died down, the idea of building this designed monument, unfortunately, was abandoned. D'Aronco was directed to other projects and continued to serve the Sublime Porte during Sultan Abdülhamid era.

It is not known for sure why this planned monument was not constructed. It is not unlikely that the unpretentious bearing of Ottoman sultans, who lived in places relatively smaller than the royal palaces found in Europe, had a role to play. The same lack of pretentiousness was most likely also exhibited after the victory won.

The project was published in Italian art magazines in Europe at the time it was prepared but it was referred to as the "Mosque in Yeni Mahalle." Raimondo D'Aronco had this corrected in a letter he wrote in 1903 to his friend Del Puppo and emphasized that the project was not of a bicycle course or a mosque, but rather a Victory Monument.[135] The silver mockup of the project submitted to Sultan Abdülhamid Han by D'Aranco is on display today at Yıldız Palace.

Years later, D'Aronco would once again design a monument project, parts of which he personally had Sultan Abdülhamid build. The monument, which was constructed in 1905, is still standing in Marja Square in Damascus. It was erected in honor of a telegraph sent by Sultan Abdülhamid II when the railroad reached Damascus. D'Aronco used railroad and telegraph elements in the same design and added a mockup of Yıldız Hamidiye Mosque, where the Sultan used to go for his Friday Prayer Procession (Cuma Selamlığı) to the top of the monument. Sultan Abdülhamid II, who was a good carpenter, did the woodworking on the mosque reproduction.

The inscription on the memorial of the Hejaz Railroad Telegraph Line reads: "This is the splendid memorial of the telegraph line, which is extended to the Blessed Hejaz Railroad by the command and decree of the Chief of Believers, the Caliph of the Universal Prophet, His Imperial Majesty and Grandeur, the Son of Sultan, the Ghazi Sultan, our master Abdülhamid Khan II."[136]

In addition, Marja Square in Damascus is not the only place where a monument was erected in memory of the Hejaz Railroad. The Great Column was constructed in 1905 across from the Haifa Train Station.[137]

40 - Another memorial project of D'Aronco

134 *Udine Çağdaş Sanat Galerisi Koleksiyonunundan Raimondo d'Aronco'nun Türkiye yılları (1893-1909)*, İstanbul 1995, Italian Cultural Center, p. 56

135 *Ibid.*

136 *Hicaz Demiryolu'nun 100. yılı*, İBB, 2008 İstanbul, p.19

137 *Ibid.*, p. 40

Sütun-u Âli - Haifa - 1905

Lagari, Hezarfen, İbrahim Efendi and Ali Ağa: Four Names, Four Incredible Projects

Lagari Hasan Çelebi attempted the first human rocket, Hezarfen Ahmed Çelebi successfully completed a flight from the Galata Tower to Üsküdar in the Murad IV era with wings that he designed himself, Head architect İbrahim Efendi brought a submarine project to life for the first time in the Golden Horn and Ali Ağa introduced rockets called "pumps" that were his own invention in his work called the Ümmü'l-Gaza fi Tedbiri'l-Harb ve Levazıma. These four incredible projects were limited to trials and their designers and the names of the projects were not mentioned again in historical sources.

According to Evliya Çelebi's *Seyahatname* (Book of Travels), in 1632, Sultan Murad Han IV had a daughter whose name was Kaya Sultan and a grand festival was organized in the Topkapı Palace to celebrate this.

At the festival, many artists and scientists who were top in their fields were gathered together and each portrayed their talents. Among these scientists and artists was a Turkish inventor nicknamed Lagari (Lean) with the name of Hasan Çelebi.

The festival began with the setup of the Sultan's throne on the waterfront. After the palace community gathered and the Muslim people were handed gifts, "this Lean Hasan invented a seven armed rocket from fifty oka (one oka equals 1.2 kg) gunpowder paste," and while eyes were on him proceeded to the presence of the worldly sultan along with his assistants. Then among the surprised expressions of the İstanbul people Lagari Hasan Çelebi mounted the rocket over the sea. Then he first saluted Murad Han then on his signal " his students lit the rocket fuse" and after a loud noise Lagari Hasan was seen to be rising rapidly towards the sky. As Lagari Hasan Çelebi was advancing into the clouds "would light the rockets next to him as well" to accelerate even more and the sparkling light given off by the rockets lit the whole sea from beginning to end. After this amazing performance when the "large rocket no longer had gunpowder," as Lagari Hasan was falling towards "the ground" everyone began to look at each other with an expression asking "What will happen now?" when right at this moment Hasan Çelebi suddenly "opened the wings that he had with him" and began to glide through the sky gradually making his way down. Finally "he fell into the sea in front of the Sinan Pasha pavilion" and removed his wet outer clothing before he came forward and "kissed the ground in the presence of the sultan" and completed this amazing "invention" successfully to win the adoration of the sultan with this talent. Sultan Murad Han IV gave a gift of "a pouch of gold" to Lagari.

Evliya Çelebi, who personally witnessed the event, completes the writing about the first "human rocket" in history after Lagari Hasan Çelebi spends some more time at the side of Murad Han IV having won his great interest and favor; with the line "After going to Selamet Giray Han in the Crimea was deceased there."[138]

It appears that the information conveyed by Evliya Çelebi was met with great interest by the science community and Lagari went down in history as the first person to attempt "a human journey to space."

The Weekly World News magazine published in America has written that the first human space flight in the world was attempted by an Ottoman Turk named Hasan Çelebi based on the Nordic Aviation Museum Director Roffavik. According to the magazine, Hasan Çelebi fired up the two story rocket that functioned with gunpowder in 1633 and the rocket reached 2.5 kilometers before falling into the sea. Space Research Department Director Frank Winter from the famous Smithsonian Institute in America also verified this information and said "The Turkish rocket man Hasan Çelebi's attempt in 1633 is the first flight with a human on record."[139]

138 Evliya Çelebi, "Seyâhatnâme," v. 1, Topkapı Palace Museum Library, Baghdad, nr.: 304, vr. 216b

139 Kaplan, Sefa, *Hürriyet*, 17.05.2010

Hezarfen Ahmed Çelebi's project and attempt is also of a nature to go down in history. Hezarfen made a successful flight from the Galata Tower to Üsküdar. The period is the Sultan Murad IV Period.

According to Hezarfen Ahmed Çelebi, before conducting the flight, he made several shorts flights in the sky "over the Okmeydanı pulpit, in star wind forces with eagle wings eight-nine times," and by practicing the drill, he determined the mechanism that would allow him to stay in the sky the longest. When the preparations for the big flight were completed, he went to "the Sultan Murad Han in Sarayburnu and while Sinan Pasha was resting in his manor, climbed to the highest point of the, Galata Tower" and quickly began flapping his wings and "Ahmed Çelebi flew with the southern wind" successfully passing to the other side of the İstanbul Strait; and "fell at the Doğancılar Square in Üsküdar."

Obviously quite impressed by Hezarfen Ahmed Çelebi's great invention "Murad Han, rewarded Hezarfen Ahmed Çelebi with a pouch of gold" to give him well earned recognition.[140]

It is clear that these wild attempts encouraged İbrahim Efendi and Ali Ağa so that the Ottomans could again witness some firsts on Ottoman soil.

Head Architect İbrahim Efendi is one of the interesting profiles of the Tulip Period. However he was not noted in history for architecture but for the development of the first submarine. We encounter this project both in Mehmed Hazini's and Vehbi's *Surname* (Event Journals).[141]

According to what is described in Seyyid Vehbi's *Surname-i Hümayun* (Imperial Festival Book); the largest of interesting events that attract interest on the fourteenth day of the circumcision festivals that were held by Sultan Ahmed III in 1719 is former architect İbrahim Efendi's "copy of a crocodile" that he built himself on his own amazing talents.

This first submarine in the world that was a copy of a crocodile at times would dive into the depths of the sea while other times rising above the sea and then disappearing from sight. In this manner the object came "slowly towards the direction of the coastal palace" and to the presence of the great sultan where he was sitting, proceeded to move for "about half an hour" and covered a large distance under the sea. But at this point it

41 - Lagari Hasan Çelebi's first manned rocket attempt

disappeared and did not resurface for a very long time. The İstanbul public could not believe what they were seeing because at that time "the extent of this art" was beyond reality to them. They were concentrating on watching the spectacle without taking a breath and were fixed in their spots with even being able to wonder. The ability to be able to succeed in expertly displaying even the dream of this difficult seeming big invention, let alone the reality in this period was enough to comprehend the greatness of this expert architect and his invention!

When the crocodile that had taken a dive into the deep waters of the Sea of Marmara resurfaced the "dancers" that came out first started dancing while clapping their hands; then they disappeared into a portion of the sea again along with the crocodile.

140 Evliya Çelebi, "Seyâhatnâme," v. 1, Topkapı Palace Museum Library, Baghdad, nr.: 304, vr. 216b

141 Seyyid Hüseyin Vehbî, "Sûrnâme'-i Hümâyûn," Ali Emîrî, History, nr.: 344, vr. 184a,186b, 187a

42 - Hezarfen Ahmed Çelebi's first manned flight attempt

Apparently İbrahim Efendi waterproofed the bottom of the boat-shaped submarine that had an outer appearance of a fish with tar and replaced the windows that he had opened in the middle with tent likenesses to prevent water from seeping in and also he had invented various weight tools connected to this attractive vessel that would enable the body to immerse in the water and resurface.

While constructing this submarine he did not neglect to foresee precautions that would quickly launch the vessel that was under the sea to the surface that were "ready for action" and made sure that these were prepared to be able to be used when the "time" came. Also on the outside, in order to be able to breath, "five-ten propeller like tools pushed up from the water" to form an aspirator for an internal ventilation system. The head architect thought that this system should not look unattractive on the vessel and manufactured to look like birds landing on the vessel on the portions of the vessel that stuck out of the water.

If it is considered that the small submarine made by British Engineer Day in 1774 was tried and a larger submarine was built in which he perished without ever reaching the water surface, it is apparent what a big achievement the Ottomans made in this topic as well. It is obvious that the Head Architect was unable to expand his project in later years, or else his life span was not sufficient to accomplish this.

Another accomplished scientist of the Tulip Period is Ali Ağa Bombardier Mine Foreman. Ali Ağa describes weapons and tools of his own inventions in his works called *Ümmü'l-Gaza fi Tedbiri'l-Harb and Levazıma.*[142] Undoubtedly, the most interesting of these are the rockets. Initially, in the work Ali Ağa explains the duties he had during the wars that were fought in this era; later he describes the cannon and artillery types and what precautions need to be taken. One of the weapons that Ali Ağa invents is rockets called "pumps," which are used in invading fortresses. In his work he describes the rockets for which he has included pictures in great detail. These are around 7-8 meters in height and can barely be encircled by a human. Ali Ağa says "I have invented a pump because the pump used by our bombardiers can only disperse fire but this pump disperses fire and with the five or ten each guns that are included in them, can also fire bullets" to explain how much more effective his inventions are compared to the old weapons. According to Ali Ağa the new rocket he has invented should especially be used in fortress invasions.

142 Ali Ağa, *Ümmü'l-Gaza fi Tedbiri'l-Harb ve Levazıma,* Topkapı Palace Museum Library, B 368

43 – The rockets invented by Ali Ağa, which he called *tulumba* and made public in his work entitled *Ümmü'l-Gaza fi Tedbiri'l-Harb ve Levazıma*

It is unclear whether or not this weapon that was designed in the years when the Ottomans had not yet lost their military strength was used. What is interesting, however, is that the drawings of the rockets are at a level that matches exactly the rockets of our present day.

An adventurer flying from the Galata Tower to Üsküdar, a passionate scientist that propels himself into the sky with fireworks, a brave head architect that can remain underwater for minutes with his team, and a clever expert that designed a rocket to be used in fortress invasions. These four Ottoman entrepreneurs with their four incredible projects were the first of their kind and continue to remain in the minds of many…

River and SAP (Southeast Anatolia Project) *Layihas* (Projects)

During the early years of his sultanate, Sultan Abdülhamid II asked governors to provide him with reports on the general situation and problems of their regions. In compliance with this request, *Layihas* (projects) began being sent from all over the country to the capital. In addition, supervisory boards were frequently set up to solve emerging problems and to monitor practices; reports were obtained from these boards. Intellectuals of the day were not deaf to the problems of the regions in which they lived; dozens of different projects were prepared and submitted to the Sultan. One of these was the *layiha* (project) Hasan Fehmi Pasha, who was an Ottoman vizier, proposed on 24 April 1880.[143]

His *layiha*, "Anatolian Public Works," was certified by Mr. Galan, the Director of Turuk-u Maabir (The School of Roads and Bridges). It was submitted to the Sadrazam (Grand Vizier) Köse Raif Pasha, sent to the relevant commission, approved by the Ministry of Public Works, discussed in the Council of Ministers, where a decision regarding it was made, and on 7 Muharrem (the first month of the Islamic calendar) 1300 AH, it was announced in *Vakit* newspaper.[144]

Hasan Fehmi Pasha was born in 1836 in Batum, went to school in İstanbul, served as parliamentarian in the first term of the Meclis-i Mebusan (The Parliament), and was one of the important statesmen of the Abdülhamid II era. He served as governor of Aydın and Salonica, London Ambassador, Minister of Justice and Professor of Law.

In his *layiha*, Hasan Fehmi Pasha proposed building highways and railways between certain centers on Ottoman territory, constructing docks and harbors, draining swamps and opening up the land obtained to agriculture, and revitalizing thousands of decares of land through irrigation canals made by setting up levees on rivers. The Pasha's *layiha* was well received by government officials. Serious steps were taken to implement it and parts of it were actualized in subsequent years. However, some of the articles didn't make it past the planning stage. The Southeast Anatolia Irrigation Project was one of them.

The Pasha argued that by draining the swamps in the region, using the Tigris and Euphrates Rivers, and providing public works and transportation services, there would be a phenomenal growth in production; and even this project could alter the image of the Devlet-i Aliye (the Sublime Ottoman State).

Sultan Abdülhamid Han II said in his memoirs, which have come down to us today, "...If we could create a rational irrigation system by setting up levees in front of Tigris and Euphrates Rivers, we would be able to transforms areas that are very dry today to the paradise they were thousands of years ago." By saying this, he was indicating that the project had been accepted at the highest level of the state and that the will necessary to implement it was available.[145]

To summarize, the *layiha* began to be implemented in Sultan Abdülhamid Han's era, with thousands of roadways and railways being constructed as part of the project. Moreover, after it was published in *Servet-i Fünun* on 7 January 1909, the pasha implemented many irrigation projects, such as the Konya Plain Irrigation Project. It would take another 100 years, however, for SAP (Southeastern Anatolia Project).

There was another irrigation project whose name does not appear among the irrigation projects, such as SAP, carried out in the country. The name of the proposal was the Kızılırmak (Red River) Irrigation Project, which was prepared by Ahmed Bey, a former director of a mine.

143 BOA, YEE, Dosya no. 11, Gömlek. 5

144 Dinçer, Celal, "Osmanlı Vezirlerinden Hasan Fehmi Paşa'nın Anadolu'nun Bayındırlık İşlerine Dair Hazırladığı Layiha" TTK, *Belgeler*, Ankara, 1971, vol: V-VIII, issue 9-12, p.188

145 Sultan Abdülhamid, *Siyasi Hatıralarım*,' Dergah, İstanbul, 1999, p. 64

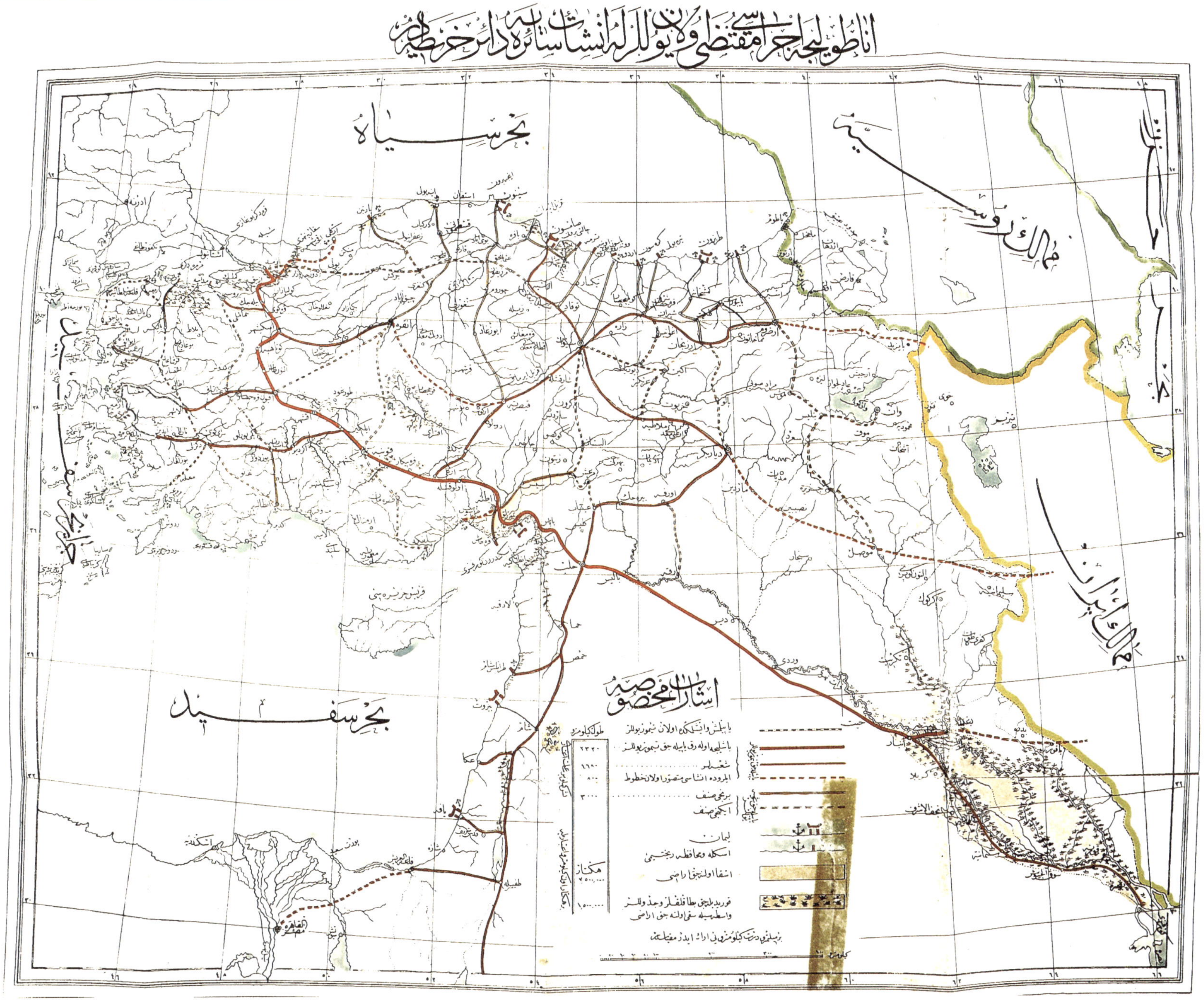

44 - The blueprint for the project prepared by Hasan Fehmi Pasha - "On Public Works in Anatolia" - 1880

Kızılırmak (Red River) Project

The individual preparing this project was an Ottoman bureaucrat but it was developed forty years before Hasan Fehmi Pasha's proposal. Sources show that it was 1848, when Sultan Abdülmecid Han was in power. Former Director of the Gümüşkan Mine, Ahmed Bey, submitted a proposal to Süleyman Pasha, Minister of Commerce, on 30 September 1848. The six-article proposal called for facilitating transportation and shipping by operating ferries on the Kızılırmak (Red) River; irrigating agricultural land by building dams, procuring lumber for the Ottoman Naval Fleet from the forests around the river; and planting olive trees, in particular, in some of the regions. The blueprint of the proposal was drawn on a 112 x 53 cm-map.[146]

Ahmed Bey emphasized in the proposal that in the event that the job of balancing the river water levels was given to him, he would do everything in his power to serve the Sultan without harming the poor, either, and provided an assurance that he expected no remuneration.

Ahmed Bey asserted that this project would make river transportation possible from the Black Sea to Kayseri and from Kayseri to Sivas. He also stated that the revenues generated by the project would readily cover expenditures, and that the project would bring five thousand kese (one kese [pouch] was 500 kurus then) annually to the treasury.[147] It is not known why this proposal was not realized but a similar one was considered for the Menderes River, too.

45 - Kızılırmak Project of Ahmed Bey, former director of the Gümüşkan Mine - 1848

Opening the Menderes River to Traffic Project

The main objective of the project was draining the swamps in the Great Menderes Plain and expanding productive activities there. The work to be done on the Menderes River, along the Bahr-ı Sefid-Aydın-Sarayköy would make river shipping possible. The project would open the Menderes River to river traffic extending 380 km.

The oldest documents proposing transportation on the river are in old İzmir English Consulate Reports. In the reports, dated 1868, 1869, 1872, 1873, 1874 and 1882, the physical structure of the Gediz and Menderes Rivers and the harm flooding caused to agriculture were recorded in the finest detail. The insistence of the consuls who wrote these reports on opening up the Menderes River to traffic is remarkable.[148]

The idea of shipping on the Menderes River was first mentioned in a consulate report dated 1869. An actual project based on the idea was proposed to the government on 5 August 1989, at which time a petition for a concession to carry out the project was also submitted. Approval of the project and the awarding of the concession to Ser Kilari (Chief Clerk of the Palace Pantry) Osman Efendi followed. The matter was looked into and it was seen that Osman Efendi had deposited in Osmanlı Bank the 1,000-Lira guarantee fund required of those seeking the concession and procured the equipment needed to carry out the said project. The contract and technical specifications concluded with him, along with *layiha*s, were submitted by the Ministry of Public Works to the Prime Minister. When the 35-article agreement and a 28-article technical specifications, organized into five sections, this project was obviously enormous and did not only encompass simple river traffic; on the contrary, it aimed at changing the relations of production for the entire area.

While it is not known why the contract concluded with Osman Efendi was not followed through, ten years later, in 1892, it was brought up again. The concession for traffic on the Menderes River was, this time, given to Ohannes Samancı Efendi and Abdullah Efendi, who were Ottoman citizens. However, because the persons making the request failed to come up with sufficient capital and difficulty was experienced setting up a company, the project was postponed to a later date.[149]

146 BOA, HRT. 0144

147 "Bir Osmanlı Maden Müdürü'nün Kızılırmak Projesi," *Yedi Kıta*, August 2009 supplement.

148 *İngiliz Konsolosluk Raporlarına Göre İzmir Ticareti (1864-1914)*. Izmir Chamber of Commerce, Izmir, 1998. The reports with said dates p.16, 20, 27, 35, 42, 47

149 Başaran, Mehmet, "Büyük Menderes Nehri Efsaneleri ve 19. Yüzyılda Nehir Ulaşımı Projesi," *Toplumsal Tarih*, Issue: 54 (June 1998), p. 50-51, BOA, Yıldız, Resmi Maruzat, File No 55, Jacket 9

46 - Architect Kemaleddin Bey's Berlin Hamidiye Hospital Project - 1901

Sultan Abdülhamid Han's Berlin Hospital Project in Appreciation for the German Fountain

In 1827, at the Battle of Navarin, an Ottoman armada was destroyed by an allied force of British, French and Russian naval vessels. A few years later, they recognized Greece, which had broken away from the Ottoman State; and in 1830, France occupied Algeria. These developments compelled the Ottoman State to seek new alliances. By the 1850s, Germany, which had been an agrarian country, had experienced the industrial revolution. It expanded its designs internationally and entered the playing field of big states. first significant rapprochement between the two states, which had been in con since the era of Selim III, took place at this time. However, the Crimean War for the Ottoman State to reconsider. This time, England and France were Ottoman all

In subsequent years, Ottoman administrators would also do what was necessar maintain a balanced policy vis-à-vis the Russian threat. Meanwhile, the enorm

ELEVATION

ارتفاع

loans contracted through the Crimean War skyrocketed with the new Russian War in 1877. In the final quarter of the 19th century, the Ottoman State experienced both economic and military difficulties; so it saw no harm in developing better military and commercial relations with Germany, which had expanded its commercial and economic goals without losing territorial integrity. Commercial activity between the two countries accelerated after Sultan Abdülhamid came to power.

Germany, making no demands for Ottoman territory, could have used Sultan Abdülhamid's political astuteness as a bargaining chip. The existence of England and France at this time could not be denied, but Germany would always take the side of the Ottomans as a third proposal that would lower the price.

Germany's share in Ottoman foreign trade began increasing in the 1880s and would continue to do so until World War II. Between 1895 and 1914, German investments in Ottoman territories would more than double, from 138 million Sterling to 282 million Sterling. 25% of foreign investments were made by Germans.[150]

This commercial rapprochement resulted in an increase in the German population, particularly in the capital. An assortment of organizations began operating in İstanbul to meet the various needs of German merchants, artisans, engineers, doctors, soldiers serving in the Ottoman Army, and embassy members and their families. The German Hospital and the German High School are just two of them. What's more, in 1896, newspapers like the *Konstantinopler Handelsblatt*, began to be published in İstanbul; everyday, 600 newspapers, including the *Konstantinopler Handelsblatt*, would be sold in İstanbul.[151]

In the meantime, the friendship between the two emperors had developed to such a degree as to require face-to-face meetings. Emperor Wilhelm visited Sultan Abdülhamid II twice—first in 1889 and then in 1898.

Wilhelm II, having left quite pleased by the second of these visits, sent a fountain made in Germany to İstanbul as a tribute to this growing friendship. This fountain would later be mounted in a place prepared for it in Sultanahmet Square. The unveiling of the German Fountain was on 27 January 1901, the birthday of Emperor Wilhelm.[152]

It would have been unfathomable for Sultan Abdülhamid II to remain indifferent to Wilhelm's courtesy . Without wasting any time, in appreciation for the German Fountain and as a symbol in Germany of Turkish-German friendship, Sultan Abdülhamid planned to have a

150 Guataert, Donald, *Osmanlı Devleti'nde Avrupa İktisadi Yayılımı ve Direniş*, Yurt, Ank ara,1987, p. 19

151 Koloğlu, Orhan, "Almanca Basın," *Dünden Bugüne İstanbul Ansiklopedisi*, TTV, İstanbul, 1994, v. I, p. 213

152 Batur, Afife, "Alman Çeşmesi," *Dünden Bugüne İstanbul Ansiklopedisi*, TTV, İstanbul, 1994, v. I, pp. 208–209

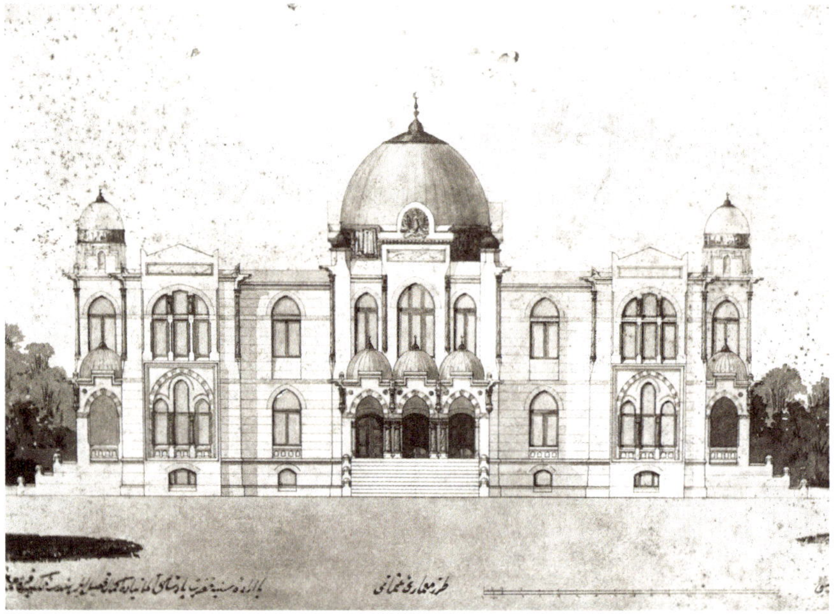

charitable institution built in Berlin for the German people. He ordered the Berlin Ambassador, Ahmet Tevfik Pasha to get the ball rolling.

At the conclusion of talks Ahmet Tevfik Pasha carried out with the pertinent persons, property was allocated on the Emperor Wilhelm Berlin University campus. The Berlin Hamidiye Hospital was to be built there using Ottoman-style architecture. The job of preparing the plan of the hospital and its cost estimation report was given to Kemaleddin Bey. Kemaleddin Bey submitted the registry and plans to the sultan at the beginning of 1902.[153]

The hospital was designed to include polyclinics and special needs, as well as 14-bed in-patient wards—two six-bed wards for men and women, and two private rooms.

Other features of the hospital to be built in Berlin by the imperial edict of the sultan were listed in its cost estimation report. The total and partial costs of the building were calculated in the registry as follows:

- Total cost of the hospital buildings: 1,310,768 kurus,
- Central building: 918,687 kurus,
- Building in the shape of a pavilion containing patient wards and an operating room: 23,335 kuṛus,
- Hospital appurtenant structures, kitchen: 42,245 kurus,
- Laundry and study building: 52,775 kurus,
- Bacteriology and chemical laboratories: 56,611kurus.

Sultan Abdülhamid II charged Vassilaki Yanko Bey, one of the imperial architects, with the job of building the Berlin Hamidiye Hospital. Yanko Bey, based information obtained from the Berlin Embassy, reported in a letter dated 14 April 1902 he sent to the Treasury that the construction would cost 20,660 Liras.

Yanko Bey was subsequently summoned to the Treasury, where, on 15 July 1902 a 32-article contract containing a payment plan for the money to be spent, along with the construction conditions, was concluded between the State Treasury and Architect Vassilaki Yanko.[154] The Berlin Ambassador, Ahmet Tevfik Pasha and a person to be appointed by Wilhelm II were to make sure that the buildings were built as desired.

Today, a building bearing the name Ida-Simon Stift and used by the Berlin Medical University Biochemistry Institute stands in the place of the Berlin Hamidiye Hospital. The Ida-Simon Foundation building, which is included in plan made in 1935 of the area where the Berlin University Gynecology-Obstetrics Clinic is located, is located in the same place as the Berlin Hamidiye Hospital does in Kemaleddin Bey's 1902 plan, except for the long north-south part of the "T"-shaped design. The Ida-Simon Foundation building was most likely built upon the foundations of the partially standing Berlin Hamidiye Hospital.

153 BOA, Y. PRK. EŞA. 38/47, 21 Haz. 1317 [15 June 1901]

154 BOA. (YEE), No: 406. Old number: YEE. No: 12, No: 151/5, Envelop No: 151, Box No: I

The Berlin Hamidiye Hospital project prepared by Architect Kemaleddin Bey shows that, after returning from Germany, he began a new architectural style he called *Tarz-ı Mimari-i Osmani* (Ottoman Architectural Style). This architectural style would later become transformed into the First National Architectural Movement through the contributions of the architect Vedat Bey.

The first work of Architect Kemaleddin, whose picture is on today's 20-Lira bank notes and who is the pioneer of the national architecture movement, was the Berlin Hamidiye Hospital.[155]

Years later, the Berlin Hamidiye Hospital plans were taken off the dusty shelves of the Ottoman Archives and, prior to the visit to Turkey made by the German President Christian Wulff made between 18–20 October 2010, were sent to Prime Minister Recep Tayyip Erdoğan by capable archivists there. Erdoğan, who has implemented many of the incredible projects of Sultan Abdülhamid, such as the undersea channel, metro and funiculars, showed the plans to President Wulff in order to transform a dream of Sultan Abdülhamid's into a reality, and said to him "we want to build." According to newspaper reports, Wulff was unable to conceal his surprise and said that "he would see about it."[156]

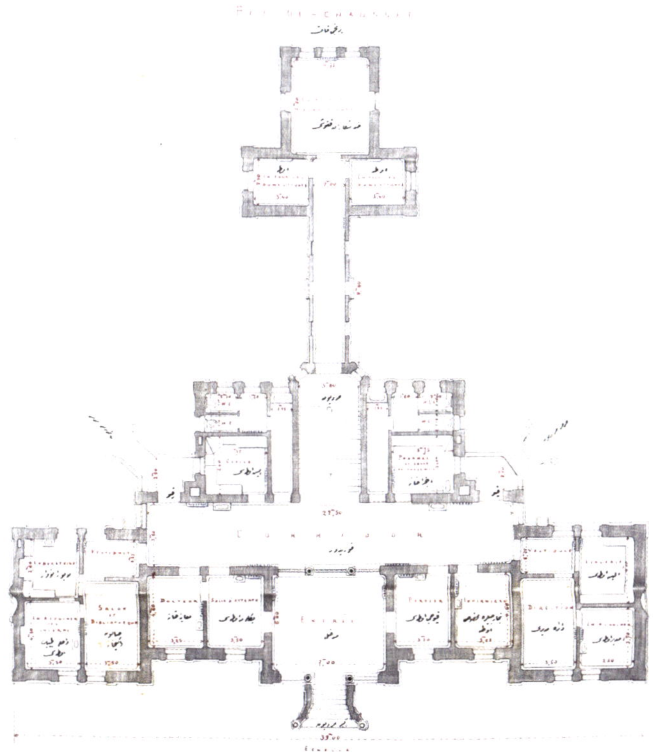

155 Yıldırım, Nuran, "Birinci Ulusal Mimarlık Akımının İlk Örneklerinden Kemaleddin Bey'in Berlin Hamidiye Hastanesi Keşif Defteri ve Planları", *Toplumsal Tarih*, Issue 188, August 2009

156 *Hürriyet*, 20 October 2010

Sea Water Desalinization Project

157 BOA, BEO, File No: 294, Jacket No: 41

158 Aşık, Melih, *Milliyet*, 13 May 1990

The shortage of potable water, one of the world's fundamental problems, has been led humanity to seek ways of obtaining it since Antiquity. Obtaining drinking water from sea water is one of these. Distillation of sea water is one of the oldest methods for doing this. Many civilizations in Antiquity used such techniques as percolation through sand and pebbles, and boiling to make seawater potable. Boiling is one of the most frequently used methods in the Middle East and North Africa.

About 400 years ago, Sr. Richard Hawkins used a simple distiller to obtain drinking water while travelling to the American continent. With the arrival of the Age of Steam, ship engineers began making even more developed instruments. However, there is no evidence available as to whether this technique was used industrially at that time.

Desalinization became a subject of modern science through a method tried for the first time in 1748 by the renowned French physicist Jean Anine Mollet. However, except for a few small experiments, no area of application could be found for desalinization because of its high unit costs.

The method used during these years was the traditional one. As in many other fields, a number of breakthroughs would appear in Ottoman territory. In the early 19th century, seawater distillation installations would be built to get drinking water from sea water, and would operate in the cities that were the sacred trusts of the Ottomans for many long years.[157]

A newspaper column article entitled "Ottoman Test by Water"[158] quoted an individual who had experienced this interesting incident:

"After the opening of an installation converting seawater to fresh water in 1976 in Jeddah, Saudi Arabia, Nejdet Özmen (Turkish Ambassador to Saudi Arabia, between 1966 and 1971) was speaking with foreign diplomats:

He begins by saying, "This first seawater distillation installation ...," but he is interrupted by the French Ambassador: "No sir, this is not the first seawater distillation installation."

"Really? What was the first?"

The French ambassador tells him that the first one was made by the Ottomans and the Turkish ambassador is stunned. The French Ambassador later gives Mr. Özmen a book. The book is entitled, Portrait of an Arab City: Jeddah. We have reproduced the pencil drawing from that book on the page below. The picture's caption reads as follows: "This condenser is a pioneer of water distillation installations and provided modest amounts of drinking water to Jeddah for decades. This installation operated until the 1940s, when it was dismantled because water from the Fatima Valley reached Jeddah."

Additional information about the installation is also provided in the article:

This installation was built by the Ottomans toward the end of the 1800s; in other words, during the Empire's years of decline. The book offers no details about the system used, but we imagine that it was based on boiling seawater in cauldrons and then transporting the steam created in the cooled cauldrons to be distilled. As salt settled to the bottom of the hot cauldron, the steam transferred to the other cauldron would be converted to water. As can be seen in the latter picture, the water was carried to the city by horse and donkey-drawn wagons.

In addition to what Melih Aşık has written, we know from correspondence on the subject discovered in the Ottoman Archives that there were installations set up in Mecca, Jeddah and, especially in Yanbu "to obtain drinking water from seawater." Among the information obtained from the correspondence is that new-type machines (in documents, the terms "Drying Oven" and "Water Machine" are used) costing 500 Ottoman Liras were brought from England to be used at these installations, and that the former method was replaced by a "the cauldron of a distillation machine and all its equipment," which produced 50 tons of water in 24 hours.[159]

We know that it was a chronic problem of the İstanbul, which was founded during the Greek migrations between 750 and 550 years BCE. The Roman Emperors Hadrianus, Constantinus, Valens and Theodosius; the Ottoman sultans Sultan Mehmed II and Süleyman the Magnificent, and many Ottoman rulers up until Sultan Abdülhamid, were forced to take measures against this problem. Sultan Abdülaziz searched for new ways of procuring water that didn't depend on the classic methods already in use. He granted a concession in 1868 to a French company, thereby securing the establishment of the Dersaadet Anonim Su Şirketi (İstanbul Incorporated Water Company). The process was completed in 1883 and, with the pumping stations construction, İstanbul got pressurized city water. In subsequent years, other installations operating with the same system would be set up. But the search didn't end. In 1893, Sultan Abdülhamid requested "an investigation into whether fresh water could be obtained by passing an electric current through seawater, as he had heard had been done in France, whether it could be used in street fountains and, moreover, what its cost would be." Upon the orders of the Sultan, the Ottoman Embassy in Paris took steps and wrote a response, including in an attachment a report that had been prepared by Dr. Bufer; it sent the two of them to the Foreign Ministry. The Ministry then sent the report onto the Sanitation Commission, which evaluated the reports and reached a decision on 19 April 1894.[160]

In the decision, it was said, "the "hermit" method, whereby drinking water is obtained by passing an electric current through seawater has not yet been accepted in Europe. In the event that it becomes used in Europe, its application in Turkey can be considered, too."

This decision was reached at the conclusion of an experiment conducted at Lorient Naval Hospital on 26 September at 8 am. During the experiment, the Mayor of Lorient Monsieur Draui, Chairman of the City Hygiene Commission Dr. Waqule, the person assigned to preparing the report, Dr. Bufer, and the study owner, Hermilt were personally in attendance.

When it came to human health, the Ottoman capital, which witnessed many breakthroughs, was perhaps even more cautious. However, being aware of the existence of this method and ordering that it be applied as soon as possible is noteworthy.

47 - Jeddah Sea Water Purification Plant

159 BOA, İ.HUS, File No: 20, Jacket: 1311/B-68

160 BOA, Y.A.HUS, File No: 294, Jacket:41

Railroad-Mercantile Shipping Project

This is a project for which only *zeyil*s (appendices) could be found on the dusty Ottoman Archives shelves. From what could be gathered from the project, there should have been seven more explanatory drawings. The existing appendix is sufficiently large to deduce main outline of the project.

The project in the archives, recorded under the heading "Monsieur Aydes's Railway Project Dedicated to Ship Transport was accepted and approved by a science commission;[161] aims at setting "any kind of ship" on rails with an installed apparatus to be transported to a second port, as is understood from the heading.

161 BAO, Y.PRK.AZJ, File No: 7, Jacket 53

However, there is no clue in this one-page document as to how this was to be accomplished. First, two pools parallel to the railroad, with a length of three thousand feet, and have different functions in the port must be constructed side by side. Moreover, there must be large doors between the port and the pools. The process will be started by first introducing a Transfer Boat to the keel of the ship with the help of a machine. At this point, the keel of the ship will be attached to the transfer boat and the retaining sides of the ship body will be leaning on the boat. The process will be completed again with the help of machines by drawing the ship that is sitting on the transport boat in deep water to the railway on shore, which is comprised of two iron bars at five step intervals from each other.

It states in the existing document that the loading process would finish in an hour on average, that the locomotives pulling the transfer boat and the ship on it must be twice as powerful, that even 20 ton ships could be carried between two ports with this system, that the locomotive could reach a speed of 12 miles (about 20 km) and that the transport was absolutely safe.

Considering that the construction of canals is very costly, this project seems to be relatively logical; moreover, the fact that the project proposal was accepted by Ottoman administrators and examined and approved by a science commission must be taken into account. Since a commission was established to review the project, there must also have been a report, in addition to the appendix of the project and the 7 drawings mentioned in the project. We hope that these documents will be uncovered soon because the answers to such questions regarding the purpose of this method, between which ports this ship transfer was planned and whether or not a new railroad line would be needed for this project are not provided in this appendix.

It is possible that this project is included in Sultan Abdülhamid's extension of the Baghdad-Basra railroad and may be another alternative to the Suez Canal like Aqaba-the Dead Sea-Jerusalem.

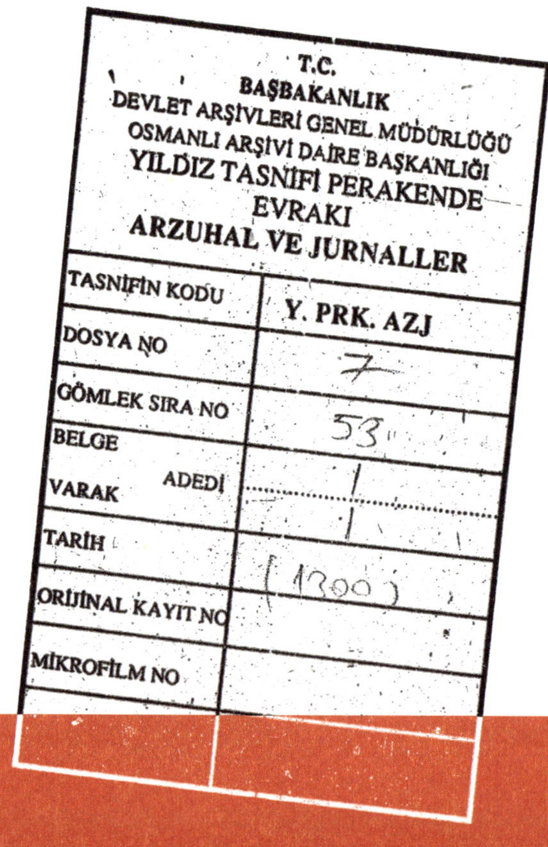

T.C.
BAŞBAKANLIK
DEVLET ARŞİVLERİ GENEL MÜDÜRLÜĞÜ
OSMANLI ARŞİVİ DAİRE BAŞKANLIĞI
YILDIZ TASNİFİ PERAKENDE EVRAKI
ARZUHAL VE JURNALLER

TASNİFİN KODU	Y. PRK. AZJ
DOSYA NO	7
GÖMLEK SIRA NO	53
BELGE ADEDİ	1
VARAK ADEDİ	1
TARİH	(1300)
ORİJİNAL KAYIT NO	
MİKROFİLM NO	

48 - Appendix to Monsieur Aydes's Ship Transport Railway Project - 1882-1883

Sample Village Project

The Sample Villages Project developed by İzmir National Library Director İsmail Bey, members of the Administration Maksutzade İbrahim Efendi, Mahmut Efendi and Doctor Şahabettin in the days when the Balkan Wars had ended and World War I was continuing, was presented to the Ministry of Internal Affairs on August 12, 1916.[162]

The Ministry examined the project and decided, "There is a need to reform the welfare and financial structure of villages for which the state is responsible; therefore, Sample Villages could be implemented under state supervision, on the proviso that they be governed with their own budget and income." However, a law needed to be passed for this (*İdare-i Nevahi Kanunu*—the Law for the Administration of Districts).

As mentioned in the presentation of the project, which carries the signatures of 4 members and one director, "Its inspiration comes from the order given by the War Office to be carried out at Military Recruiting Branches and published in newspapers for the establishment of sample villages in the four corners of the country where veteran soldiers could settle and developed further from this order. So it appears that the name "Sample Villages" came from the War Office.

The project that was prepared by the İzmir National Library director and administration members is comprised of a presentation, 42 articles under the heading of "Necessities for Sample Villages," 16 articles in answer to the question "What is Found in Sample Villages?" and 8 articles in answer to the question "Who benefits from Sample Villages?"

The reason for us starting the project explanation with "in the days when the Balkan Wars ended and World War I was still continuing" is a requirement in answer to the 8 article question "What are sample villages beneficial for?" These answers leave the impression that the project was also developed for the continuation of a generation whose population had been jeopardized by war and epidemics.

The benefit that is hoped for in the project is the development of a healthy new generation. The additional articles have the goals of increasing the population and prosperity, maintaining morals and protecting forests, increasing well-trained individuals, developing industry, education and agriculture and training good soldiers.

The requirements for this have been listed in 42 articles. However, before this, there is another heading with 4 articles. The committee is asking for authority on four subjects under this heading in order to be able to start on the project.

First of all, the selection of village to be reformed must be made. The determination of clothing (this is not clear, but it's possible a uniform was considered) for the population, property, village leader and committee of elders, preparation of marriage-birth-death-event schedules and the improvement of households. In addition, the village must be provided with springs, roads surrounded by trees, sewers, place to wash the dead, sanitary areas, coffee houses for contemplation, signs that teach etiquette and socializing, girls and boys schools, public sports facilities, a sample room, pharmacies and a certified midwife. Also police that take action according to orders, a village court of peace comprised by the village leader committee, a voluntary road guard, a simple municipality organization and a fire unit must be formed.

49 - Presentation of the Model Villages Project developed by National Library Director İsmail Bey, and Administrative Associates - 1916

162 BOA, DH.UMVM, File No: 7, Case 53, 12 Za. 1334

163 Söylemez, Haşim, *Aksiyon*, Edition, 722, October 2008

164 Konuk, Nevval, *Osmanlı Mirası*, Sam, Ankara, 2008, p. 224

In the project workshops for ironworks, carpentry, lime mines, bricks, car, blacksmiths, locksmiths, shoes, tailors and book are necessary to be established for the village population's livelihood.

The project owners are not in any need of money to put the project into practice because they believe that the improvements can be made by using what the village can provide. The library director and administration members introduce themselves as a committee in the presentation. By saying "Our committee, which considers it to be our duty to serve for the benefit of the country, would like to work as volunteers," it has been emphasized that there is no expectation for any kind of personal gain from this project. The main goal that this project should reach is explained by the statement, "Rather than building a whole new village, the selection and improvement of an existing one and the application of the experience that is required to carry out the project and enable it to be spread to other villages in the best way."

It is apparent that this project which was on the agenda in 1916, set an example to projects like Köy-Kent (Village-Town) and Köydes (Supporting the Substructure of Villages) which would be put into practice in later years. However, before this project, the villages in which "all precautions for roads, water, health, culture and education had been taken" were created for veterans and refugees. In other words, the "Public Village Project" is not just an Ottoman dream that remained at the planning stage. The Refugee and Veterans villages that were established at different points throughout the Ottoman geography after the 1897 Crete uprising were the first of these applications. These villages have another shared feature; the cost of their establishment was provided by the Treasury. Today, Hamidiye Village is one of these villages. Immigrants from Crete still live there today. It bears the name of its founder. It is located outside Ottoman territory and separated into two along the Syrian and Lebanese borders.[163] The other two villages having the name Hamidiye, where Crete immigrants were settled, are now within the borders of Greece. The structures that remained from the Hamidiye Villages that started construction in Rhodes in 1906 and İstanköy in 1908 are still existent to this day. The 103-household Hamidiye Village, which was constructed in the Rhodes Kumburnu region by master Yorgis Tsimestas from Aya Yorgi, was completed in 1907.[164]

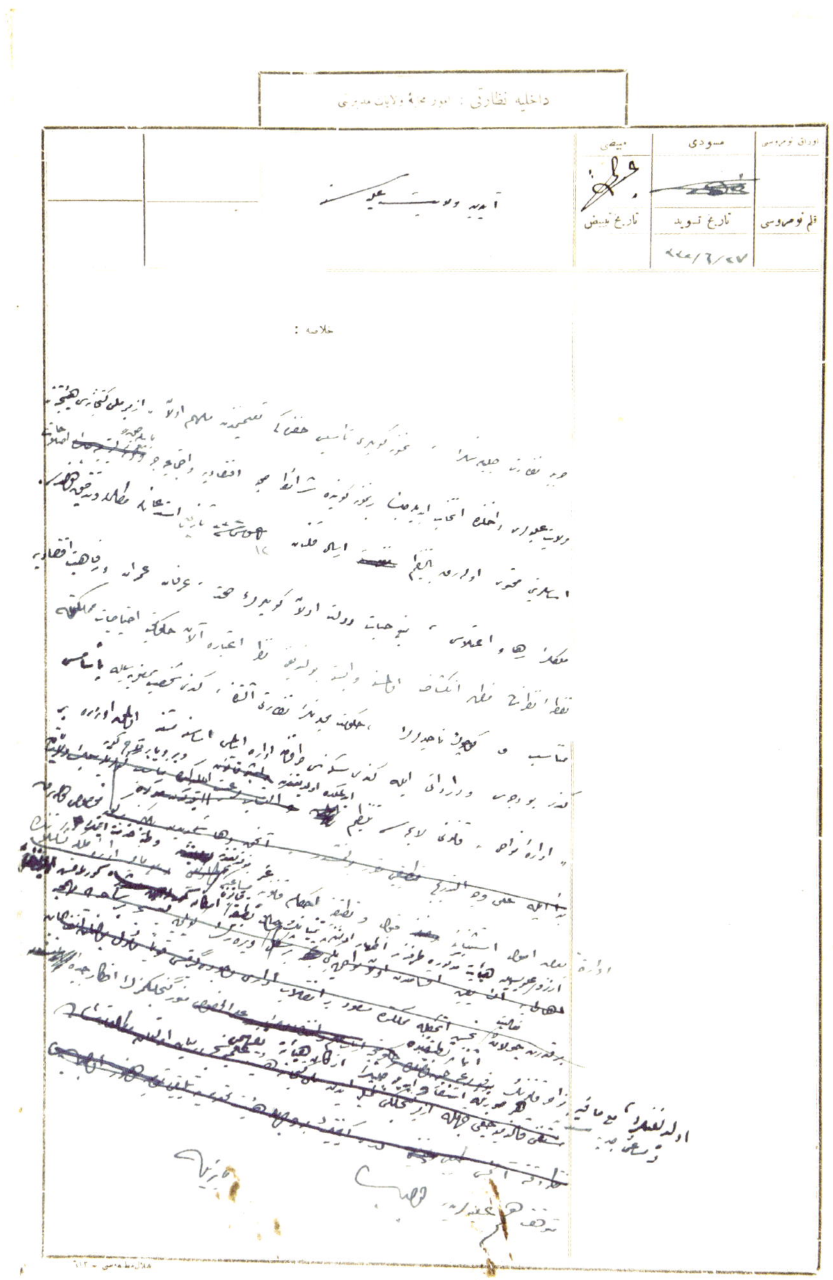

50 - Opinion of the Ministry of Interior of the project - 1916

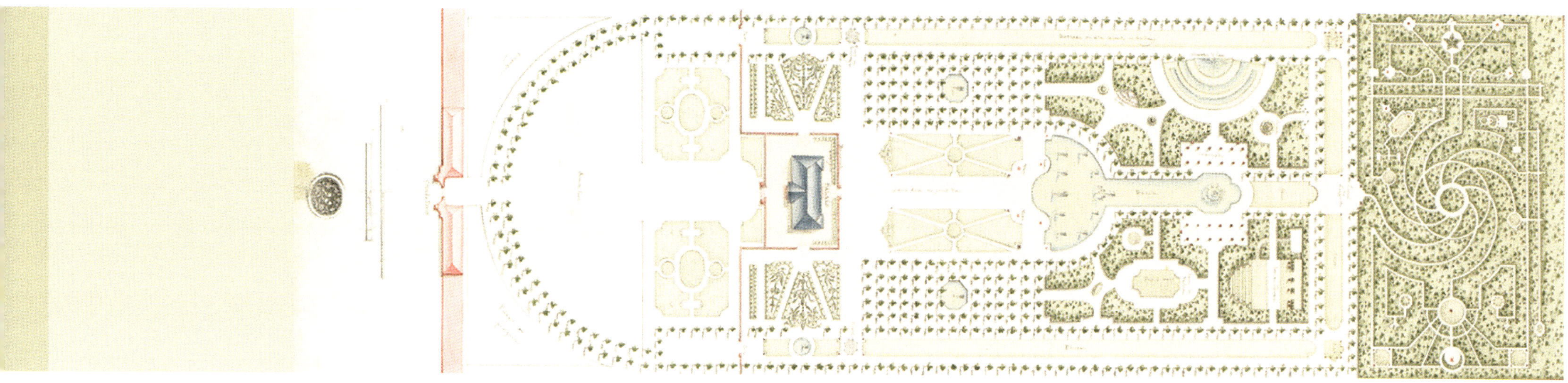

A Recreational Area Project during the Selim III Era

Palace gardens are prominent genuine Ottoman gardens for which we have little detail about their arrangement other than that their beauty is described as "the garden of Eden." These gardens are also a manifestation of the refined taste of Ottoman people. Lubenau, who visited Bursa in 1588, said, "There is a lovely garden along the wall towards the town. In the middle there is a square pool full of cut stones and in the middle of the pool on a marble column there is a lovely summer resting pavilion, also made out of marble,"[165] to express his admiration. The gardens of the Edirne Palace and the Old and New Palace that were built later in İstanbul are able to produce in admirers the same kinds of feelings. *Risale-i Beşir Çelebi* describes the Edirne Palace in his work with the same admiration, "The architect has constructed a great pavilion and palace. He has surrounded the whole area with such vines and fruit and vegetable gardens to liken it to a paradise that the beauty cannot be described with words."[166]

This state, which was specific to palace gardens, was reflected on coastal palaces that were built starting with the Sultan Selim III era and later the gardens of the coastal palaces were transformed into areas of public recreation. In fact Sultan Selim, who lived in Topkapı Palace, spent most of his time at the coastal palaces of his sisters in Eyüp and Boğaziçi. Despite that era not reaching our day, it was an area of coastal palaces in terms of architecture. The Capitol had literally turned into a private garden of the Sultan.[167]

First of all Selim III had the semi-stone Şevketiye Pavilion built on the fortress walls in front of a Baroque garden right next to Mahmud I's Topkapusu Coastal Palace. Selim III's sisters and nieces, such as Beyhan, Hatice, Şah, Esma and Hibetullah, built their coastal palaces, which were reflected in the engravings of Melling, Allom and Fladin in this era as well. The numerous foreign ambassadors in the Ottoman capital were also among those that had summer homes built along the coastal strip that extended from Tarabya to the Black Sea. Also strong pashas of the era like Sadrazam Koca Yusuf Pasha took their place in residences along the Bosphorus.[168]

The coastal palaces, pavilions, seaside residences and summer house were not only in competition with their architecture but also with their gardens. It appears that this application that was an important part of the Ottoman city culture became the fashion of the era with the entry of this application into the area of interest belonging to a branch of art. When the descriptions offered by Yirmisekiz Mehmed Çelebi Efendi on his return

165 Atasoy, Nurhan, *Osmanlı Has Bahçeleri*, Ministry of Culture and Tourism Publications, İstanbul, 2005, p. 89

166 Atasoy, Nurhan, *ibid.*, p. 90

167 Esemenli, Deniz, *Bir Reformcu, Şair ve Müzisyen: Sultan III. Selim Han*, Greater Municipality of İstanbul and Ministry of Culture and Tourism Publications, İstanbul, 2008, pp. 31–35

168 Esemenli, Deniz, *op cit.*, pp. 35–41

from Paris are added, the situation departed from being a personal interest and the government became officially involved in the subject.

At the beginning of the 18th century, Yirmisekiz Mehmed Çelebi Efendi was sent to France to conclude an agreement with France, but he returned without making a new agreement with the child king Louis XV. In the meantime, he had the opportunity to stay in Paris. When he arrives at the office, rather than the agreement, Yirmisekiz Mehmed Çelebi Efendi has in his hands.[169]

Also written in great detail in this legation journal are the palaces, palace gardens and amusement areas of Paris. When Çelebi is describing the Cloud Palace he says: "… and there was one pool that we saw on a slope from the top towards the bottom in which stairs had been woven with marble. As the water flows the steps are covered to form the image of a staircase of water. They have placed sprinklers in various places and in the mouths of dragons that are a surprising sight to see," being unable to conceal his admiration. "We arrived at the spectacular heartwarming palace called Marli and the spectacle of its lovely garden. Such a fancy recreation area has been built that there can be no match. I think the organization style of its garden should be chosen above all" were the comments of Yirmisekiz Mehmed Çelebi Efendi, which undoubtedly attracted a lot of attention and curiosity in the Ottoman palaces. In fact, after Mehmed Efendi's return, Damad İbrahim Pasha changed the bed of the Kağıthane River, had an eight-hundred- meter portion of the river covered in marble and a pavilion that sits on thirty columns built, in front of the pools of the pavilion, artificial waterfalls, dragons with water pouring out of their mouths and various sprinklers were placed. The visual integration was completed with the addition of many colorful mansions and pavilions to this recreation area that people could not get their fill of looking at.[170]

The transformation that started upon the return of Yirmisekiz Mehmed Çelebi Efendi was certainly not just built on his descriptions. Mehmed Efendi also brought with him documents that were used for the purpose of obtaining ideas and samples on the construction and decoration activities of the French Palace. This collection in the Topkapı Palace archives, which includes architectural information, grew very quickly due to the interest staying fresh in later periods.[171]

The idea of a fixed and constantly mobile entertainment complex in a new style for the public, which was repaired and transformed after being made in the Tulip Period and demolished by the Patrona Halil Uprising in 1730, became a current topic again in later periods along with large garden compositions and coastal palaces that were literally sprinkled throughout İstanbul.

As can be seen by the crescent and star figure in the drawing that is the subject of our work, the samples that were thought of for Paris were not just copied identically but most likely the famous names of the period were given orders for the work.[172]

The existence of such recreation areas can be explained by public demand. The changing Ottoman individual is now in need of leaving the courtyard surrounding the house that they share with their harem and intends to expand into wider circles. In order to do this there is a need for shared areas. As explained, Kağıthane Garden, which had fallen into a state of needing repair as the result of an uprising, was quickly restored along with additions and presented for recreation due to the same demand. Without further delay, the Ottoman people would show themselves in Göksu, Beykoz and Çamlıca. The recreation areas of the 18th century and this plan, with its immense expanse, have maintained their basic values and are at the same time proof of socialization. These areas, which were also frequented by foreigners, were Ottoman solutions to Beyoğlu and Pera. As in the Kağıthane area, the pavilion and coastal palace gardens (other than the walls that surrounded the structure) are areas open to the public and are each an element of this solution.

This incredible project we have on hand has been drawn with glazed tile, colored with water color, expands in parallel towards the front and back face around a small pavilion (even though the phrase "Palais" is used in the project), has an Ottoman seal and is a "Recreation Area" plan dated 1203 AH. A courtyard was planned as the prominent element immediately on the palace or pavilion surroundings and the doors that open to the outer garden from four sides gives the impression of being open to the public.

Such large recreation areas with their pools, waterfalls and crescent and stars woven among spiral flower beds were at the top of the service list of administrators at that time. The unknown details about the project, on which tiny French notes were written, were not limited to architectural elements. For whom and where it was to be applied are also important questions.

169 Rado Şevket, *Yirmisekiz Mehmed Çelebi'nin Fransa Sefaretnamesi*, Hayat Printing house, İstanbul, 1970

170 "Sadabad," *Dünden Bugüne İstanbul Ansiklopedisi*, v. 6, p. 385-386

171 Irepoğlu, Gül, *Thoughts on Western Sources in the Topkapı Palace Museum Treasury Library*, Topkapı Palace Museum Yearly I, İstanbul 1986, p.618

172 TSM Archive, H 1881

Selim Sırrı Bey's "Sports Holiday" Proposal

Selim Sırrı Bey's (Tarcan) proposal was set forth to be traditionalized on April 29, 1916 and is a project that was realized at the Kadıköy Union Sports Club Lawn. The following year, the scope was expanded and organized in the same place on May 11, 1917 and then because of the unfavorable conditions of the period was put on a shelf and remained there until May 19, 1939.

In Selim Sırrı Bey's letter of request submitted to the Ministry of National Education before this application dated March 27, 1916,

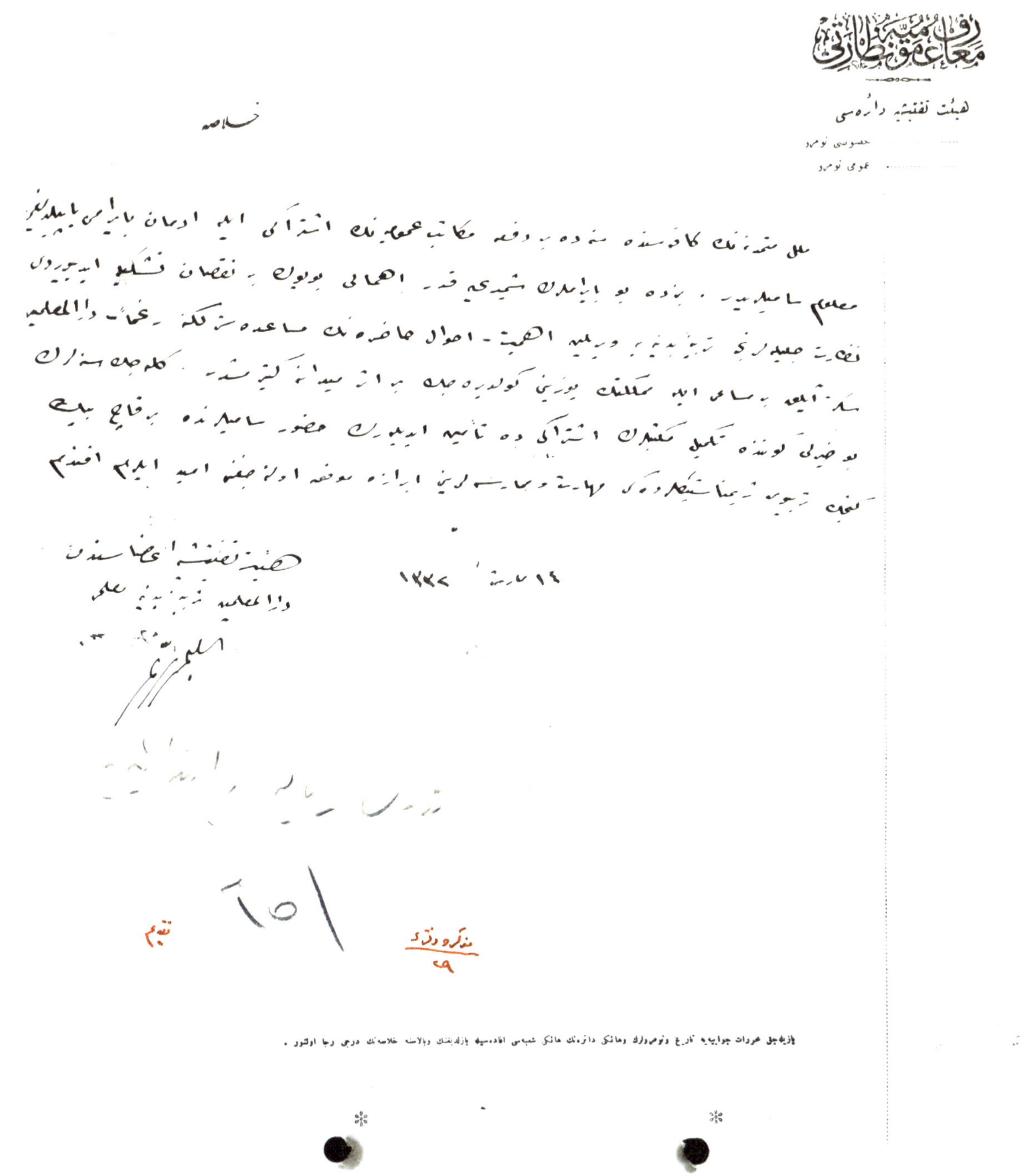
معارف عمومیه نظارتی

هیئت تفتیشیه دائره‌سی

خصوصی نومرو

عمومی نومرو

خلاصه

51 - Selim Sırrı Bey's "Sports Holiday" proposal submitted to the Ministry of Interior - 1916

173 BOA, MF, TLY, 514/51

174 Hür, Ayşe, "Geleneğin İcadı ve 19 Mayıs Bayramı," *Taraf*, May 18, 2008

175 Yurdsever Ateş, Nevin, "19 Mayıs nasıl bayram oldu". Toplumsal Tarih, P. 113, May 2003, p.34-37

176 Hür, Ayşe, *ibid.*

he emphasized that "Despite the Ministry not giving its consent, the Darülmuallim (Teacher's School) worked for eight months to create a work that put smiles on the faces of the country's people" and requested that this beneficial work be allowed in future years with the participation of other schools.[173] In this request, an event that was organized in the Darülmuallim without permission for the first time is being described. One month after this request Selim Sırrı Bey's proposal will be accepted and on May 29, 1916 the first comprehensive "sports holiday" will be realized on May 29, 1916.

Selim Sırrı provides appendices to the mentioned letter of request that include ideas for the sports holiday he want to have organized. The first appendix is comprised of a 9-article proposal and the second is comprised of a 12-article program.

In the first article of his sports holiday proposal, Selim Sırrı Bey says "The first will be organized on 1332 May's first Friday at the Kadıköy United Sports Club by the Darülmuallim-i Aliye and in following years other schools will join in participation" to give place and time information and in further articles has included information concerning the other matters involved. The Sports holiday will include sports disciplines and gymnastics performances. It will be comprised of gymnastics, body movements, school games and short distance races. Students between the ages of 12 and 16 will participate in gymnastics. Individuals older than 16 will participate in the walking races, jumps, pentathlon, tug of war, javelin throw, bicycle, pipe throw, shooting, handball and hurdle races and the winners will be rewarded. Girls and Boys schools will be invited to the sports holiday, each school will be assigned to previously determined locations and no athletes other than the students will be accepted in the races.

In the event that the suggestion is accepted, the holiday program will start with "a speech about the Sports Holiday and the Olympic games by the Darulmuallim Sports Discipline Master Selim Sırrı Bey" and continue with performances by the Darulmuallim Students Gymnastics Team including conducting a sample class from the school performances and a portion of the students will invent a game to jump from a wooden horse over the desks, mace and shot put throwing, pentathlon, high jump, one hundred meter dash, relay races, bicycle races and finally tug of war.

On the other hand, after the races are finished a football game will be played between two teams of Darulmuallim students, which will start at 14:30 and finish at 16:00.

In the final sentences of the request, it is mentioned that a "public ferry will be operating to bring students to watch the games and then take them back."[174]

This request was accepted and the first sports holiday, as mentioned above, was held on April 29, 1916 at the lawns of the United Sport's Club in Kadıköy. 200 male students participated in the performances from Darülmuallimin-i Aliye, by first performing samples of their gymnastic moves in their white shirts, black pants and red belts and then proceeded to do moves on gymnastic equipment. In the second sports holiday, held on May 11, 1917, there was a greater variety in the gymnastics competitions and, as specified in Selim Sırrı Bey's request, high jump, long jump, pole vault, javelin throw, disk throw, tug of war, 100-meter dash, 800-meter endurance running and bicycle races were organized. Also drills were conducted to teach how injured and unconscious people should be handled and awards were given to the winners of the competitions.[175]

The Sports Holiday, which was neglected from 1917 until 1928, was to be celebrated after this under such names as Sports of Discipline Festival, Schools Holiday, Gymnastics Festival, Atatürk Sports Holiday and on June 20, 1938 a clause was added to "Regulation 2739 on religious and official holidays" in order for the 19th of May to become the official "Youth and Sports Holiday."[176]

Mustafapaşalı Osman Nuri's 3,500-HP Machine Design

From what can be understood of the explanations made on Mustafapaşalı Osman Nuri's proposal he is "a second grade mechanical engineering student."[177]

Our development scale in engineering is not just comprised of a science and academic institute's history. At the same time, it represents the modernization history of the Ottoman State. The Ottoman school approaches engineering from a rationalist point of view, which is different from the West. In fact the word *mühendis* (engineer) is from the root of the word for measuring meaning measurer. In the West the word for engineering is derived from the root for "creativity" which is "engineering." This word comes from "engine" and portrays an applied approach.[178]

Engineering in the modern sense in the world was freed from being an art form with French Marshall Sébastien le Prestre de Vauban (1633-1707), and presented with the opportunity to become a science. Thanks to his works, which were published after his death, the new engineering movement spread throughout all of Europe. Vauban is accepted as a pier of engineers from a certain aspect. In his work on the subjects of combat, strike and defense techniques, as he has pointed out himself, Vauban develops many engineering theories and applications in order to stop Turkish movements.[179]

The engineering theories and application that Vauban developed against Turkish movements would enter Ottoman lands which have an ancient tradition of Muslim Theological Schools, in a very short time and the books would be memorized by engineering students in the following years; in fact, they would be trained in Humbaracı Ocağı (Mortar Corps) in 1735, Mühendishane-i Bahri-i Hümayun (Imperial School of Naval Engineering) in 1773, Mühendishane-i Berri-i Hümayun (Imperial School of Land Engineering) in 1795, Hendese-i Mülkiye Mektebi (Civil Engineering School) in 1883 at a level no less than that of modern standards at the time. These students were to be understudies in this historical development period at certain institutes to such famous masters of the time Ahmed Pasha, Mehmed Said Efendi, Mustafa Ağa, Seyyid Hasan Hoca, Hüseyin Rıfkı Tamani, Hafız İshak Efendi, Yahya Naci Efendi, Hüseyin Tevfik Pasha and Ahmed Besim Efendi.[180]

Ahmed Besim Efendi became the head engineer at the Tersane-i Amire (Ottoman Shipyard in İstanbul) in 1873 upon the departure of head engineer Shanks and continued this job until 1909. During this period Ahmet Besim Pasha made designs for a steam machine. The machines that he made designs for were manufactured and installed on various ships. Although the school where he was a student is not mentioned in his proposal, it is likely that Mustafapaşalı Osman Nuri was one of his students. Osman Nuri's project which was completed as "a triple cylinder machine of three-thousand five-hundred horse power" can be found among the National Palaces Collection. Although its placement gives the impression of being presented to the sultan this is not sufficient to arrive at a definite conclusion. However, the triple cylinder 3,500 horsepower machine project of the second year student should be taken seriously in terms of being a manifestation of a very old tradition. This is because the Ottoman State is not just among the advanced countries for the ship industry. The factories in which these students work are in competition with the French in the airplane industry. In fact, the Ottoman State would be entering the War of Independence with locally made airplanes.[181]

Later on, the first automobile and truck factory would be established with 450 workers and 48 trucks would be produced in a very short time.[182]

177 National Palace Portraits Collection. Inv. no. 12/2720

178 M. Kaçar; A. Bir, "Mühendishane'den İstanbul Teknik Üniversitesi'ne kısa mühendislik tarihimiz," *Bilgiye Yatırım Birlikteliğin Gücü*, İTÜ Mustafa İnan Library, İstanbul Technical University Rectorship, Number 1639, İstanbul, 2008, pp.14–33

179 Beydilli, Kemal, *Mühendishâne*, Eren, İstanbul, 1995, p. 181–183, 202-212

180 Kaçar, Mustafa, "Osmanlı Devleti'nde Mühendishanelerin Kuruluşlu ve Bilim ve Eğitim Anlayışındaki Değişmeler," İstanbul University Social Science Institute, Science History Department Master Thesis, İstanbul, 1996, (Collected from the Whole Work)

181 Şimşek, Muhittin, *Yarım Kalan Devrim Rüyası*, Alfa, İstanbul, 2006, p.109

182 Azcanlı, Ahmet, *Türk Otomobil Sanayinin Tarihsel Gelişimi*, OSD, İstanbul, 1995, p. 57

52 - Mustafapaşalı Osman Nuri's 3,500-HP, 3-cylinder engine project

Decision of the Meclis-i Vükala (Council of Ministers), dated 17 June 1918, regarding the Construction of a Tunnel or Bridge between the Asian and European Continents

Yes, you have read the date right—June 1918. In six months, the Çanakkale Straits would be passed, İstanbul would be occupied and on November 7, 1918 neither the council of ministers that issued this decision nor the Public Works Authority that submitted this idea into the agenda of the council would remain in existence. Minister of Public Works Münif Bey would not flee to Germany like the head of the government Talat Pasha but would move to Anatolia to take a place among those that organized the War of Independence. Today we must consider that this was a decision made correctly when the end had been reached while waiting for the results of a government that had lost its first world war.

On December 19, 1917 with Ministry of Public Works letter no. 52, work was started with the belief that "the provision of a connection and access with such tools as a tunnel or bridge between the European and Asian Continents would entail a lot of benefit for the country." 15,000 Liras are needed for the purchase of tools that will be used in the survey and building.

The Ministry of Public Works has indicated that the supply of this money could be possible with "the appropriations for railroad and ports surveys and construction of the year 334" and in order for this work to be started quickly in the coming April, it is stated in the letter that an agreement needed to be made with "the commission executing a detailed investigation of the Yenikapı Port."

The letter is read in the June 17, 1918-dated Council of Ministers meeting and the proposal was evaluated before a decision was made. In the Council of Ministers decision, it is stated that the arrangement of the agreement would be possible with "the addition of a special

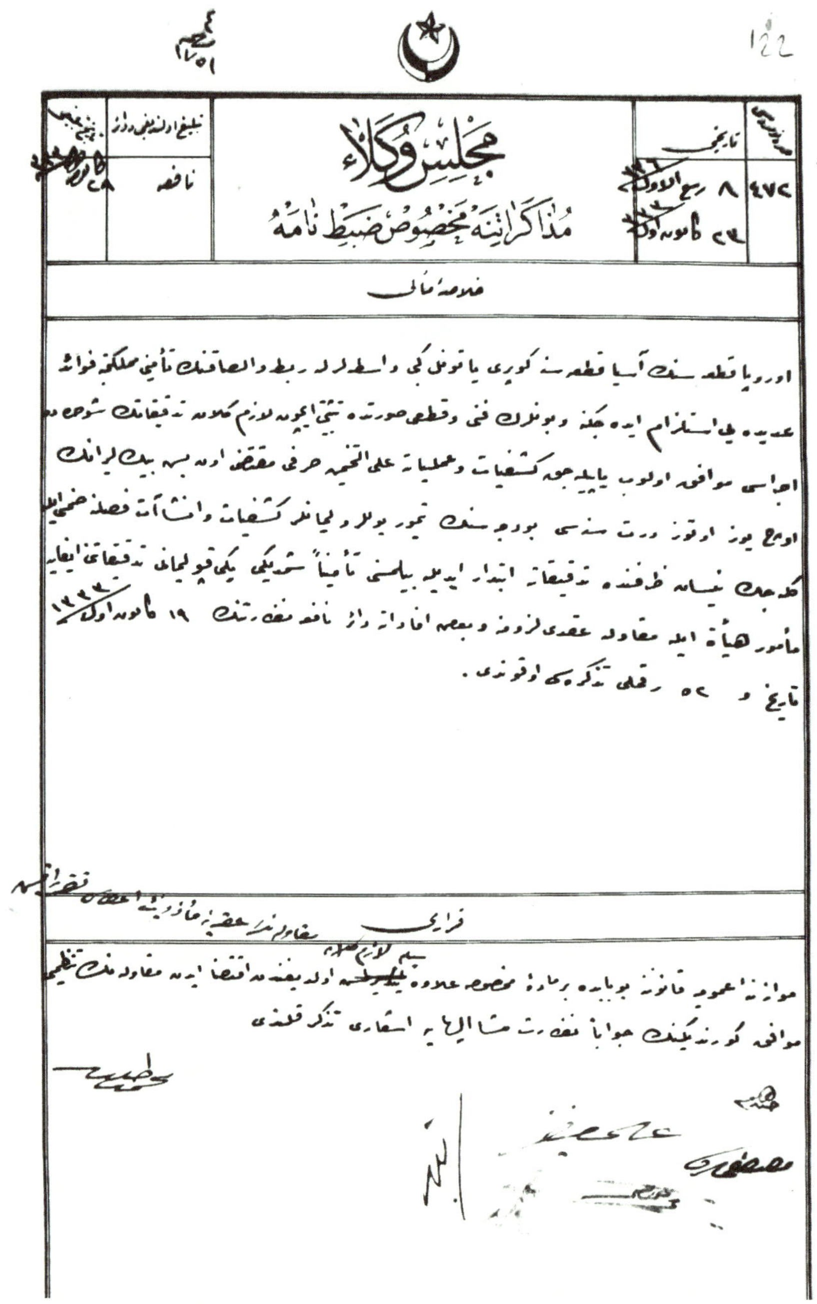

مجلس وکلا

مذاکراتنه مخصوص ضبط نامه

article to this effect in the Public Regulation" and in fact this article is seen to be appropriate. Later, the way is paved for the enterprise with the words, "in response to the arrangement of the agreement being seen as suitable the mentioned has been negotiated."[183]

Before the project is entered into the agenda of the Council Public Works Minister of the period Ali Münif Bey and Maison Zsigmondy Béla representing the Austria-Hungary joint venture sign a ten-article agreement on April 11, 1918.[184]

In the letters before the agreement despite the phrase "underground between İstanbul and Üsküdar" being used a second alternative project after the Council of Ministers decision becomes the subject of signature. The agreement that is signed with Münif Pasha does not include tube passage, but rather "An iron fixture bridge on stone foundation."[185]

The fact that Simon Préault's 1890 proposal is referred to in the Agreement caused the text of Préault's proposal to be archived in the same file as Zsigmondy Béla's proposal.

The fact that despite 28 years having been passed since Simon Préault's proposal and 16 since Strom's proposal for a Tunnel Passage the idea has stayed fresh in the minds of Ottoman administrators no doubt shows that the desire to eliminate one of the difficulties of the İstanbul geography never completely faded away. What is in the essence of the Great Ottoman Consideration is that no matter what the conditions are, to bring any kind of problem into the agenda and create solutions. It is very important that the idea for the Tunnel Passage and Bosphorus Bridge of today had its groundwork laid in some of the Ottoman State's most difficult days.

183 BOA, MV, File No. 210, Jacket No. 12, 08.Ra.1336

184 BOA, File No. 34T370, Fund Code. 230, Loc. 95.40.2/19-21

185 BOA, File No. 34T370, Fund Code. 230, Loc. 95.40.2/4-16

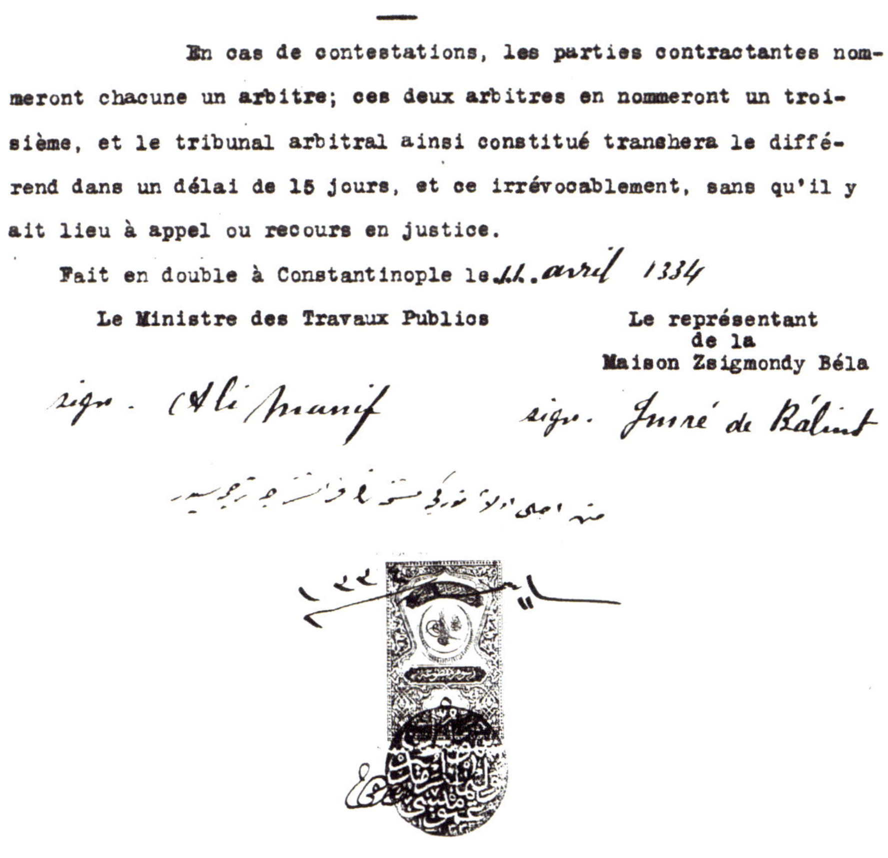

En cas de contestations, les parties contractantes nommeront chacune un arbitre; ces deux arbitres en nommeront un troisième, et le tribunal arbitral ainsi constitué tranchera le différend dans un délai de 15 jours, et ce irrévocablement, sans qu'il y ait lieu à appel ou recours en justice.

Fait en double à Constantinople le 11. avril 1334

Le Ministre des Travaux Publics

sign. Ali Munif

Le représentant de la Maison Zsigmondy Béla

sign. Imré de Bálint

Yıldız Garden and Railroad Project

Yıldız Garden, which is known to have been forest in the Byzantine Era, had been used as hunting grounds for Sultan Süleyman the Magnificent. In the Ahmed I period, it was included in the Hadak-i Hassa or Sultan's gardens. The first building of the area was also constructed in this period. Structures in Yıldız Garden, where Murad IV also hunted frequently, increased rapidly in the Selim III, Mahmud II and Sultan Abdülmecid periods. Sultan Abdülhamid II actually left Dolmabahçe Palace and settled into Yıldız. Now Yıldız had become a Sultan's residence. In this period, both Yıldız and the property grew in conjunction with the building.

In the new period, the Little Mabeyn, Cariyeler Manor, Şale Manor, Yıldız Mosque, harem buildings, a theatre, carpenter, pharmacy, repair shop, locksmith, porcelain workshop, library, ammunition workshop, sultan's sons sections, pavilions, administrative buildings, guard and service buildings were added to this large social complex.[186]

Yıldız had now taken on the old Topkapı atmosphere of the Classical Ottoman Era and had literally been transformed into a small town with an internal palace population of 12,000. When Ertuğrul, who was responsible for the palace's security, and the soldiers from the Orhaniye base, whose numbers reached 20 thousand, were added, Yıldız was to become a place that was in need of social supplies.[187]

A large garden with water canals passing through it to fulfill the Sultan's area of interest, a zoo with a rich array of animals, as well as a lemon orchard had been built; however, there was no area in this small but disorganized town of nearly 32 thousand people that could be used as an organized recreation area. This huge area that was surrounded by the Yıldız and Orhaniye Mosques and Ertuğrul Dervish Lodge was also comprised of building units that were disconnected from each other. This was the reason for another difficulty that frequently came up. A plan was prepared to provide transportation between the structures and for the creation of a joint recreation area.[188]

According to the plan, the garden would be expanded towards the Bosphorus and Big Mabeyn, Çadır, Malta and Şale Manors would be connected to each other with a railroad that extended about two kilometers. This railroad would also connect the Ertuğrul and Orhaniye barracks.

There were several different arguments about the garden, which widened towards the Çırağan Palace. A large pool with an island in the middle was going to be built in front of the Çadır Manor, a waterfall that gradually expanded over steps between the two hillsides, two small pools to the north of the Malta Palace and a stylish bridge to provide passage for the railway through the middle of the Çadır Manor pool as specified in the plans, was going to be constructed. The second bridge in the plan was referred to the "Big Bridge" and had 17 arches. The garden arrangement example that seemed to be separate immediately behind the Şale Manor was not prevalent over the whole project. It is probable that the land being on an incline did not allow for this. The plan did not include the whole green area between Yıldız and Çırağan Palace. It had been prepared taking the north faces of the Şale and Malta Manors as a basis. The portion of this face with a view over the Bosphorus had been left outside of the plan.

186 Batur, Afife, "Yıldız Sarayı," *Dünden Bugüne İstanbul Ansiklopedisi*, vol. 6, pp. 520–526

187 Candemir, Murat, *Yıldız'da Kaos ve Tasviye Kitabı*, İlgi, İstanbul, 2007, p. 36

188 İstanbul University Rare Works Library, No. 93283

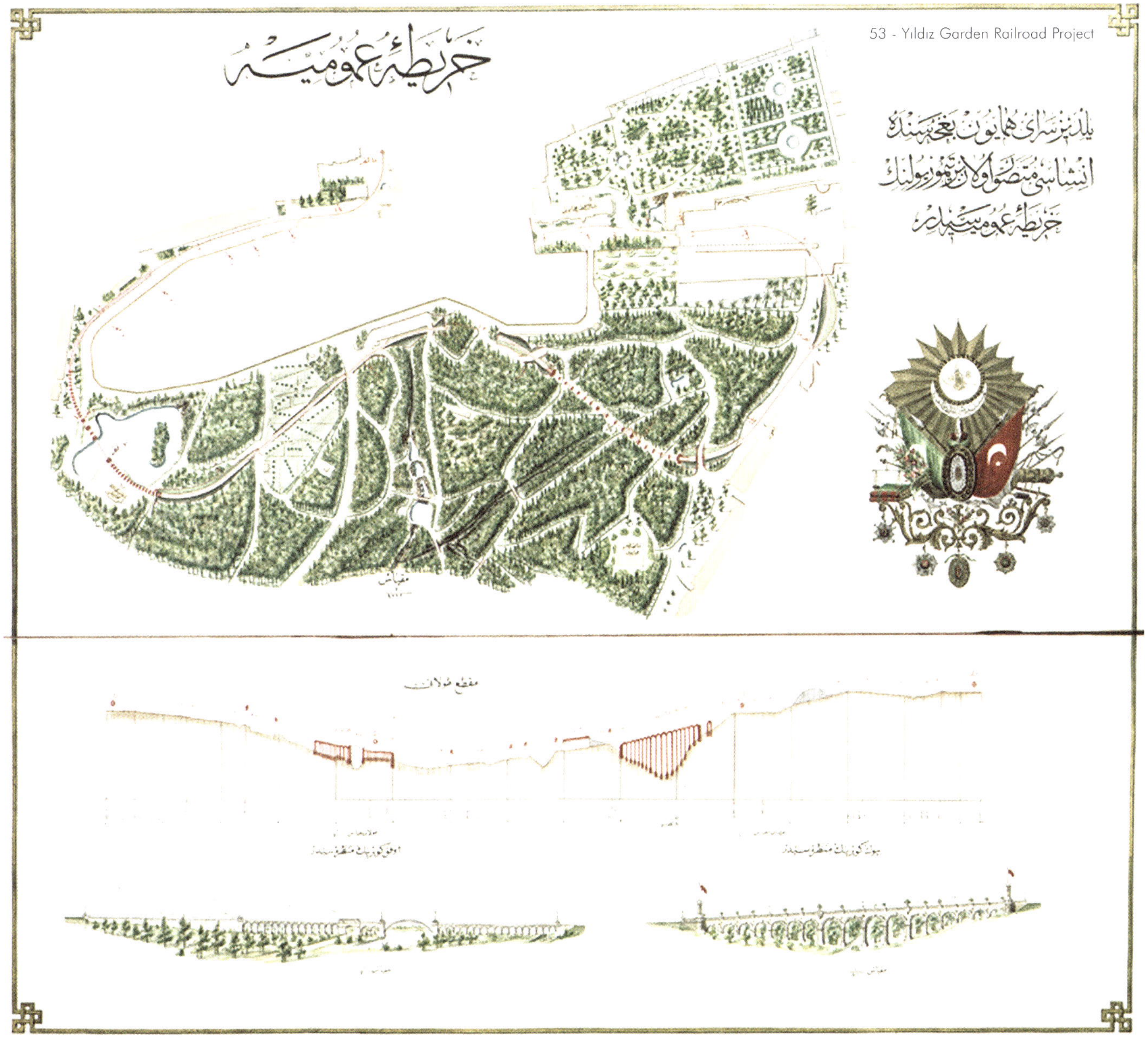

53 - Yıldız Garden Railroad Project

Yıldız Garden is now one of the largest park areas inside the İstanbul city center and hosts thousands of visitors every day. However, buildings that are disconnected from each other can still be seen. Also this disconnection is not just present between the structures. There is also a variety in the organizations that the structures are affiliated with from an administrative standpoint. The Palace is affiliated with the Ministry of Culture, some of the manors with the National Palaces Authority, some manors the Greater İstanbul Municipality and some with the private sector. Also there are buildings of other country organizations present. Yet, as seen in the plan, İstanbul is very much in need of a nostalgic recreation area of which its parts can be accessed easily.

54 - Aram Tahtaciyan's Galata Tower Project, which appeared in *Servet-i Fünun* - 1908

189 BOA, File No. 3056, Jacket. 48, Fund Code. ŞD. – NF. MRF.

190 "Galata Kulesi", *Dünden Bugüne İstanbul Ansiklopedisi*, v.3, p. 361

191 *Servet-i Fünun* (anonymous text), 29 Kânunusani 1324, p. 924

Aram Tahtaciyan's Galata Tower Proposal

Aram Tahtaciyan submitted his request to the Prime Ministry on November 9, 1908 with the last sentence, "I ask that the construction of the towers complete wood portion with various metal additions like hanging gardens, terraces and loggias be assigned to me."[189] Aram Tahtaciyan especially emphasizes the fact that he is not making any kind of financial demand for the building in his proposal. Also he is committing to leaving 20% of the operation profits to the government. What he is asking for is to build-operate-hand over for a period of 30 years.

In the years that this proposal was submitted to the prime ministry, this kind of enterprise was not without reason. The tower's only wood portion, the cone at the top had been exposed to many years of fire and storms and in 1875 the cone was removed and replaced with the construction of two polygonal pavilion rooms one above the other with a flag at the point where they ended.[190] In other words, the unique design was departed from for the first time and the demolition of the fortresses that surrounded the tower as well as the appeal of the new center of attraction gave hope to entrepreneurs. One of these entrepreneurs, as mentioned, had presented himself in his presentations and in some newspaper articles as Ecole de Beaux-Arts graduate architect—engineer Aram Tahtaciyan.

In the request submitted by Tahtaciyan to the prime ministry, the project was not very detailed; "hanging gardens, terraces and loggias with additions" were mentioned. More information was included in the article published in *Servet-i Fünun* before the prime ministry decision.[191] Also in parenthesis, this article is very significant in terms of entrepreneur-press relations. It is obvious that the article was directed at affecting the ministry's decision. The article that spoke very highly of the project and its designer is concluded with the implied question, "Let's see, what answer will be given to the project" In fact this question is asked in quotes.

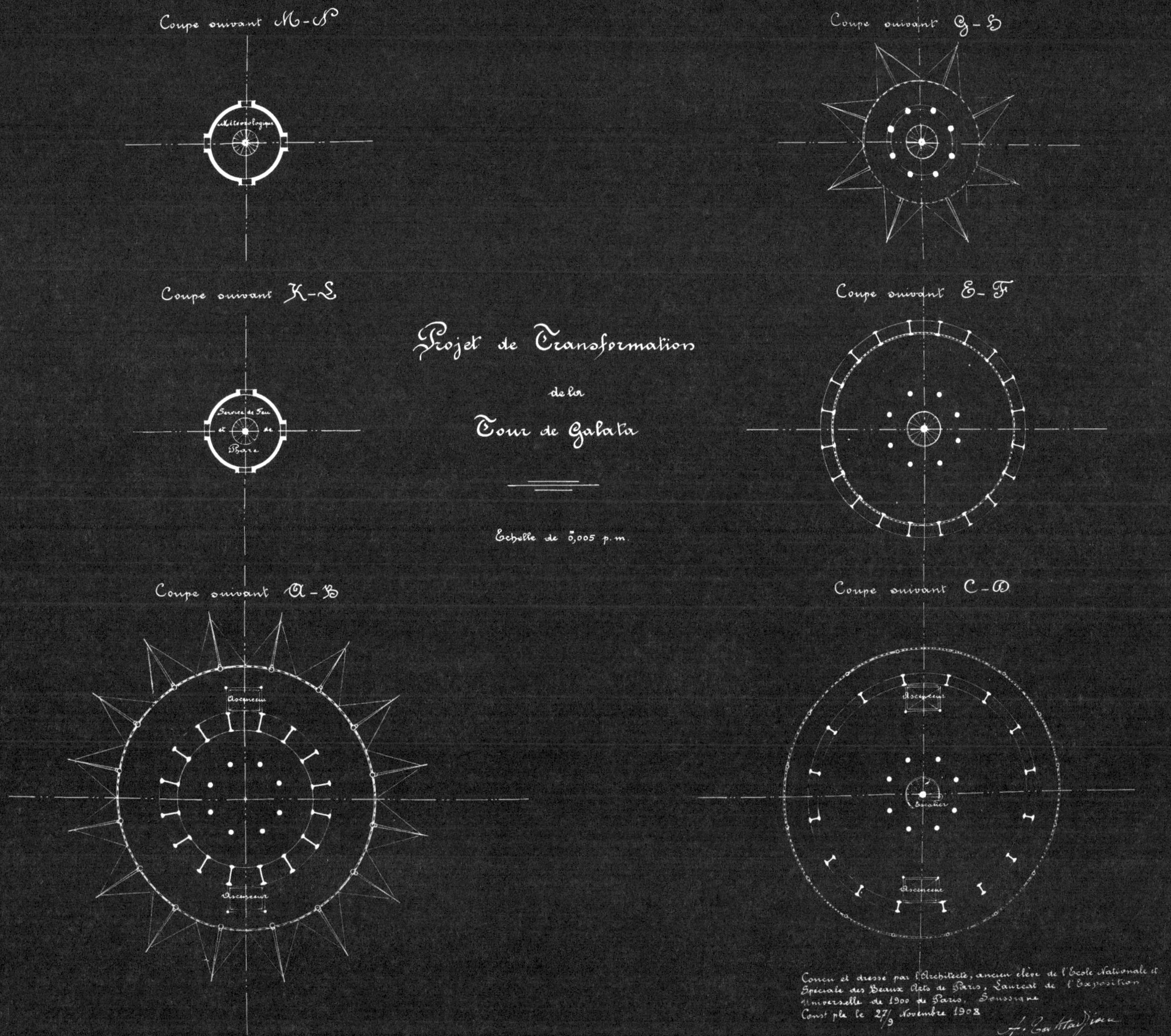

55 -The only drawing of the Project in the Prime Ministry's Ottoman Archive - 1908

192 BOA, File No. 2798, Jacket. 62, Fund Code, DH. MKT.

According to the article,

The pavilion sections that were added in the 1875 restoration will be removed, the tower will be worked on starting from the level of the old cone and a metal construction reaching a height of 40 meters would be added. This new section was comprised of a cylindrical structure on the bottom with two conjoined sections with a globe-like shape would provide an area inside that was seven or eight times the original. This new addition would be reached from the ground floor by two big elevators. These floors were to be arranged between the hanging gardens to include an excellent theatre, large restaurant and several excursion rooms.

The project also had features with a public benefit basis. In the highest point a small observatory "equipped with the best tools," a telephone and telegraph center to inform authorities about fires in the city, a time globe to measure time correctly and four giant projectors "to shed light on the four corners of the city at night" were other elements of the project. Also on the top floor "the best telescopes for enjoying the Bosphorus, Marmara and Haliç scenery" would be placed for the enjoyment of visitors.

Two months after this article, Aram Efendi's proposal was submitted to the Ministry and evaluated on April 18, 1909.[192] The decision that was written five days after the March 31st incident, despite including some concerns, was relatively positive. In the decision text which referred to article eight of the Asar-ı Atika (Old Monuments) Regulation, it was written: "it is obvious that the wood portion of the tower is not deemed in accordance with Asar-ı Atika Regulation and in its existing state is quite dilapidated; therefore, it will in any case become more attractive with construction and additions than its current state" to pave the way for this enterprise.

After Aram Tahtaciyan, who in the article said, "Not a cent need be spent by the Treasury. Let me be the one to spend. Allow me the profits for 30 years. Within this time let me undertake the expense of the tower's repair. Let the government be exempt from this expense. After that let the whole of the Galata Tower become the Ottoman's" received a positive response from the Prime Ministry in September of 1909. The project was again considered at the Council of Ministers[193] in January of 1910, along with consideration for various commission reports. In the end, the proposal was declined by the Council of Ministers.[194]

It is worth noting that the fact that despite the use of such phrases as "quite excellent and captivating," "thought of in the most excellent manner" and "a valuable work of art" to describe the proposal and "the engineer and architect of a city," "intelligent," "competent" and "expert" to describe Tahtaciyan in the article, the proposal was rejected, shows how little the Turkish bureaucracy of the time paid attention to the press.

193 BOA, File No. 126, Jacket. 27, Fund Code, MV.

194 BOA, File No. 1096, Jacket. 9, Fund Code, MF. MKT.

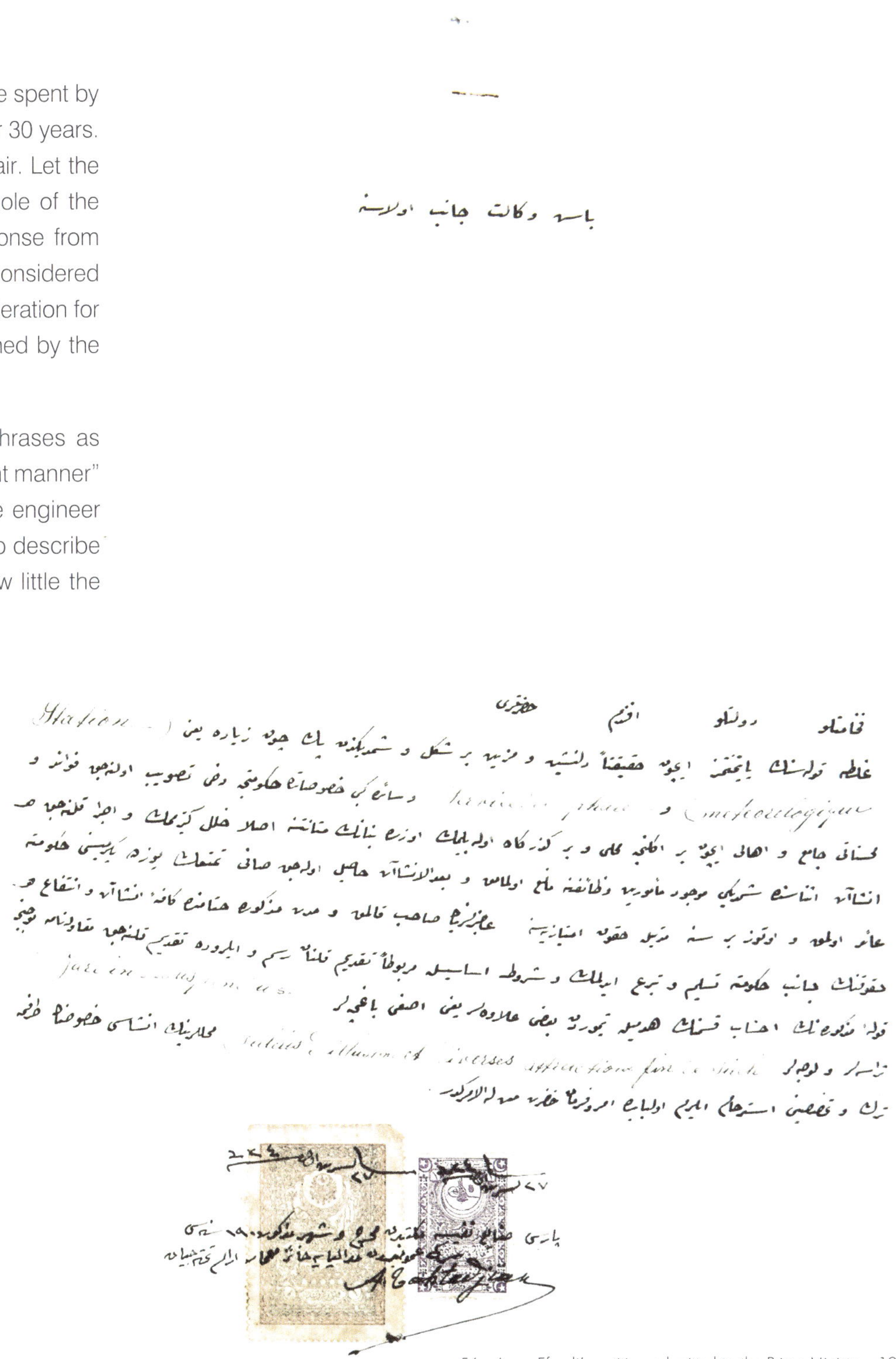

56 - Aram Efendi's petition submitted to the Prime Ministry - 1908

List of Reference for the Images

1 - İstanbul University Rare Works Library No. 90592

2 - İstanbul University Rare Works Library No. 90593

3 - İstanbul University Rare Works Library No. 90594

4 - İstanbul University Rare Works Library No. 90595

5 - İstanbul University Rare Works Library No. 90591

6 - M. Nicoletti, Architettura Liberty in Rome, Italy1978

7 - National Palaces Painting Collection, Inv. no. 12/2557

8 - İstanbul University Rare Works Library, No. 93209

9 - Gavand Eugene Henri, Tunnel de Constantinople, Paris 1876 Archeology Museum Library

10 - Gavand Eugene Henri, Tunnel de Constantinople, Paris 1876 Archeology Museum Library

11 - Prime Ministry Ottoman Archives (BOA), T. NFİ, No. 1364/22

12 - Prime Ministry Ottoman Archives (BOA), İrade-i Hususiye 1637/1146

13 - Greater İstanbul Municipality (İBB) Ataturk Library, Cat. no. Hrt.006226

14 - İBB Atatürk Library, Cat. no. Hrt. 00401-13

15 - General Directorate of State Archives, Republican Archives, Dossier no. 34T370, 230/95.40.2

16 - İBB Atatürk Library, Map Catalog, 3064

17 - İstanbul University Rare Works Library No. 93295

18 - Prime Ministry Ottoman Archives, YA.HUS, 411/174

19 - Prime Ministry Ottoman Archives, YA.HUS, 411/174

20 - Prime Ministry Ottoman Archives, YA.HUS, 411/174

21 - Ms.L.fol.66r. Paris, Bibliotheque de l'nstitut.

22 - Galleria Casa Buonarroti in Italy. No. 31250

23 - Prime Ministry Ottoman Archives, Y.PRK.TNF, Dossier No: 4, Case 21, 23. Şa. 1312

24 - İstanbul University Rare Works Library No. 824

25 - National Palaces Painting Collection, Inv. No. 12 /2800

26 - Zeynep Çelik states in her work "Değişen İstanbul" (Changing İstanbul) that the drawing is in the Prime Ministry Ottoman Archives. However, despite extensive efforts, it could not be found in the source indicated.

27 - It was formed in the direction stated in the map source. Kayra Cahit, Eski İstanbul'un Eski Haritaları (Old Maps of Old İstanbul), İstanbul: İBB,

1990, p. 66

28 - Prime Ministry Ottoman Archives, HRT.h583

29 - Prime Ministry Ottoman Archives, HR.SFR. 21/39

30 - Prime Ministry Ottoman Archives, HR.SFR. 21/39

31 - Prime Ministry Ottoman Archives, HR.SFR. 74/16

32 - Prime Ministry Ottoman Archives, ADNVS.MHM.d, 7/721

33 - Prime Ministry Ottoman Archives, HR.SFR. 74/16

34 - It was formed in the direction stated in the map source. Sultan Abdülhamid, *Siyasi Hatıralarım* (My Political Memoirs), İstanbul: Dergah, 1999, pp. 103–104

35 - Prime Ministry Ottoman Archives, Y.Prk.Tkm, 34/58 29.za.1312

36 - It was formed in the direction stated in the map source. Kayra Cahit, *Eski İstanbul'un Eski Haritaları* (Old Maps of Old İstanbul), İstanbul: İBB, 1990, p. 66

37 - GAMUD colleciton

38 - Yıldız Palace Museum

39 - GAMUD collection

40 - GAMUD collection

41 - Anonymous drawing

42 - Anonymous drawing

43 - Topkapı Palace Museum B 368, Ali Ağa Ümmü'l Gaza fi Tedbiri'l Harb ve Levazimiha

44 - Prime Ministry Ottoman Archives, Y.EE, 11/5

45 - Prime Ministry Ottoman Archives, HRT.h0144

46 - Prime Ministry Ottoman Archives, Y.EE,d, 406

47 - Melih Aşık, *Hürriyet*, 20 October 2010.

48 - Prime Ministry Ottoman Archives, Y.PRK.AZJ, 7/53

49 - Prime Ministry Ottoman Archives, DH.UMVM, 7/53

50 - Prime Ministry Ottoman Archives, DH.UMVM, 7/53

51 - Prime Ministry Ottoman Archives, MF, TLY, 514/51

52 - National Palaces Painting Collection, Inv. no. 12/2720

53 - İstanbul University Rare Works Library , No. 93283

54 - *Servet-i Fünun*, 29 Kânunusani 1324, p. 924

55 - Prime Ministry Ottoman Archives, PLK.P, 6548

56 - Prime Ministry Ottoman Archives, DH. MKT. 2798/62